# CALIFORNIA STATE MINING BUREAU

### FERRY BUILDING, SAN FRANCISCO

**FLETCHER HAMILTON**       State Mineralogist

San Francisco       December, 1918

# Mines and Mineral Resources

OF

# NEVADA COUNTY

**CHAPTERS OF STATE MINERALOGIST'S REPORT**
**BIENNIAL PERIOD 1917-1918**

CALIFORNIA STATE PRINTING OFFICE
SACRAMENTO
1919

# CONTENTS.

# ILLUSTRATIONS.

# NEVADA COUNTY.

## INTRODUCTION.

Nevada County, the banner gold producing county of the state, is from 12 to 20 miles in width and eighty miles in length, stretching from the Sacramento Valley to the state line of Nevada on the east. It is bounded on the north by Yuba and Sierra counties, on the east by the State of Nevada, on the south by Placer County, on the west by Yuba County, and contains 974 square miles or approximately 623,360 acres. From the so-called 'thermal belt' of the rolling foothills at an elevation of 400 feet, there is a rapid but gradual rise in altitude until in the eastern end of the county, the snow capped peaks of the Sierra Nevada reach an elevation of 8000 to 9000 feet. This range in elevation gives wide variations of climate and diversified agricultural products. In the western portion of the county, in the 'thermal belt,' where semi-tropical orange and lemon groves thrive, snow and frost are a rarity. In the central section of the county, where the rainfall averages 45 inches and snow seldom occurs, walnuts, berries, apples and Bartlett pears thrive. In the eastern portion, where the rainfall averages 70 inches and the snowfall averages 10 to 20 feet, lumbering, stock raising and ice cutting are the principal industries.

### Water resources and drainage.

Nevada County is specially favored with a wonderful drainage and water storage system that has already been developed by a number of water, power and mining companies, giving ample and cheap hydro-electric power and also furnishing abundant water for irrigation purposes. Over 1000 miles of ditches carry water to all parts of the county and the main reservoirs of the county have a storage capacity of forty billion gallons. The water system of this county is capable of much greater development. Nevada County is capable of supplying the Sacramento Valley with immense quantities of water for irrigation, after this water has first been used for the development of electric power.

The northern part of the county is drained by the Middle Fork of Yuba River, which divides Nevada County from Yuba and Sierra counties on the north. The South Fork of the Yuba and Deer Creek, both flowing in a westerly direction, drain the central portion, while the southern part of the county, which is bounded by Bear River, is drained by that stream and its main tributaries, Wolf and Greenhorn creeks.

2—46900

## NEVADA COUNTY—Table of Mineral Production, 1880-1918.

| Year | Copper Pounds | Copper Value | Gold, value | Granite Cubic feet | Granite Value | Lead Pounds | Lead Value | Silver, value | Stone, value | Misc. Amount, (tons) | Misc. Kind | Misc. Value |
|---|---|---|---|---|---|---|---|---|---|---|---|---|
| 1880 | | | $2,702,362 | | | | | $70,144 | | | | |
| 1881 | | | 3,700,000 | | | | | 9,500 | | | | |
| 1882 | | | 3,500,000 | | | | | 10,000 | | | | |
| 1883 | | | 3,000,000 | | | | | 8,000 | | | | |
| 1884 | | | 2,950,000 | | | | | 5,000 | | | | |
| 1885 | | | 2,577,873 | | | | | 4,836 | | | | |
| 1886 | | | 3,221,038 | | | | | 8,333 | | | | |
| 1887 | | | 2,719,374 | | | | | 2,477 | | | | |
| 1888 | | | 2,600,000 | | | | | 5,000 | | | | |
| 1889 | | | 2,249,335 | | | | | 5,633 | | | | |
| 1890 | | | 1,969,613 | | | | | 14,713 | | | | |
| 1891 | | | 2,207,986 | | | | | 14,181 | | | | |
| 1892 | | | 1,945,406 | | | | | 8,326 | | | | |
| 1893 | | | 2,067,203 | | | | | 1,229 | | | | |
| 1894 | 83,728 | $7,535 | 1,830,135 | | | | | 476 | | 290 | Mineral paint | $5,800 |
| 1895 | 83,255 | 3,325 | 1,789,816 | | | | | 400 | | 150 | Mineral paint | 2,250 |
| 1896 | 28,200 | 2,820 | 2,380,756 | | | | | 8,584 | | | | |
| 1897 | | | 1,885,251 | 1,100 | $2,200 | | | 8,116 | | | | |
| 1898 | 30,000 | 3,000 | 2,017,628 | 2,000 | 1,500 | | | 19,476 | | 50 | Mineral paint | 1,000 |
| | | | | | | | | | | 6,000 | Pyrite | 30,000 |
| 1899 | 45,438 | 7,084 | 2,171,510 | 2,000 | 1,500 | | | 17,784 | | 300 | Mineral paint | 5,400 |
| 1900 | 150,960 | 20,472 | 1,812,069 | | | | | 66,841 | | 5,400 | Pyrite | 28,620 |
| 1901 | 39,588 | 6,225 | 2,121,054 | 1,000 | 3,000 | | | 18,122 | | 2,925 | Pyrite | 17,550 |
| 1902 | 28,500 | 3,975 | 2,142,740 | 2,170 | 4,160 | | | 6,124 | | 78 | Pyrite | 429 |
| 1903 | 4,500 | 585 | 2,468,047 | 2,335 | 5,395 | | | 3,252 | | | | |
| 1904 | | | 3,130,301 | 2,155 | 2,570 | | | 9,565 | | | | |
| 1905 | | | 3,179,715 | 9,525 | 9,300 | | | 32,523 | | | Platinum | 20 |
| 1906 | | | 2,658,420 | 12,840 | 9,300 | | | 24,219 | | | | |
| 1907 | 22,062 | 4,418 | 2,162,088 | | | | | 17,505 | | | | |

| Year | | | | | | | | | | | Unapportioned, 1900-1909 |
|---|---|---|---|---|---|---|---|---|---|---|---|
| 1908 | 30,166 | 4,104 | 2,297,963 | 700 | 2,100 | 663 | $25 | 21,914 | $1,678 | | 400,000 |
| 1909 | | | 2,680,235 | 1,250 | 2,800 | | | 24,926 | 1,874 | | |
| 1910 | | | 2,533,483 | 2,225 | 3,215 | | | 16,506 | | | |
| 1911 | 1,665 | 209 | 2,199,147 | 1,250 | 3,500 | 14,881 | 667 | 15,691 | | | |
| 1912 | | | 2,061,956 | | | 1,785 | 80 | 22,830 | | | |
| 1913 | | | 2,918,783 | | | 2,090 | 92 | 26,542 | 5,000 | | Gems ........ 60 |
| 1914 | 39 | 5 | 3,301,948 | | | 145 | 6 | 27,000 | 2,108 | | Other minerals .... 1,050 |
| 1915 | 1,817 | 318 | 3,466,722 | | | 1,567 | 74 | 23,762 | 3,675 | | Chromite ...... 12,795 |
| 1916 | 3,487 | 858 | 3,669,878 | 100 | 100 | 1,038 | 71 | 35,741 | 1,225 | 981 | Other minerals[1] ... 23,475 |
| 1917 | | | 3,682,947 | | | * | * | 52,335 | 1,600 | 1,962 | Chromite ...... 43,449 |
| 1918 | 42,203 | 10,424 | 3,070,453 | | | * | * | 78,567 | 1,400 | 3,528 | {Other minerals[2] ... 47,101 |
| | | | | | | | | | | | {Chromite ...... 116,933 |
| | | | | | | | | | | | {Other minerals[3] ... 29,884 |
| Totals | 541,648 | $75,367 | $101,033,272 | 40,650 | $60,640 | 22,177 | $1,015 | $740,155 | $18,560 | | $766,816 |

[1]Includes manganese, platinum and tungsten.
[2]Includes asbestos, lead, platinum, and tungsten concentrates.
[3]Includes asbestos, lead, manganese, platinum, and tungsten concentrates.
*Under "Other minerals."

### Relief.

The western portion of the county is characterized by rolling hills of irregular outline and elevation. The central part, especially in the vicinity of Grass Valley, at an elevation of 2500', is composed of irregular plateaux. The streams before mentioned divide the county into a number of ridges separated by deep V-shaped cañons 1000' to 2500' in depth with a general trend of from east and west to northeast and southwest. Except where the streams are crowded together, these ridges are broad and have a general slope to the southwest as has the rest of the range.

Nevada City (elevation 2500') is the county seat, and had a population of 2689 in 1910. Grass Valley (elevation 2400'), pre-eminently a mining town, had 4520 inhabitants according to the census of the same year.

### History and production.

The placer mines of Nevada County have been worked from 1849, while one of the earliest worked deposits of gold-bearing quartz in California was the Gold Hill mine, Grass Valley, which was discovered in the summer of 1850. From 1849, up to and including 1915, gold to an amount variously estimated at from $238,000,000 to $275,000,000 had been taken from the gravel, drift, hydraulic and quartz mines of this county. Since 1880 the gold production has totaled over $100,000,000, coming principally from the quartz mines of the county. The table herewith (pp. 2 and 3) gives the yield of all minerals in Nevada County, by years since 1880.

Previous to 1883, in which year the hydraulic mines were restrained from dumping their tailings into the rivers and streams, the largest hydraulic mines in the world were operated in this county. As the result of a single cleanup, the Malakoff mine near North Bloomfield shipped one gold brick valued at $114,000.

### History.

Following the discovery of gold, the main body of prospectors were soon on the Bear River, Wolf and Deer creeks and in the South and Middle forks of the Yuba River. By 1850 Nevada City and Grass Valley were populous mining camps. From 1849 onward the development of the alluvial placer and the subsequently discovered 'dead rivers' or Tertiary river channels was rapid.

The auriferous quartz veins, the original source of all the gold of the Sierra Nevada, were discovered and developed and mills were erected in 1850 and 1851 on the various lodes which have since become famous producing mines. The extreme richness of these discoveries stimulated a keen interest in quartz mining and in rapid succession the following mines were discovered and worked: Grass Valley, Gold Hill (1850), Massachusetts Hill (1850), Empire (1850), North Star (1851), New York Hill (1851), Granite Hill (1851), Eureka (1851), Osborne Hill (1851), Union Hill (1850).

In the Nevada City district, the Providence and Merrifield veins were discovered and opened in 1851.

As has been stated before, owing to the crude methods of mining and metallurgical treatment, many of these ventures proved failures, causing a resultant period of depression in the industry.

In the Grass Valley district, the exodus of miners and capitalists to the Comstock and the flooding of the mines in the winter of 1861–1862 further caused the lode mining industry to suffer. From 1862, however, a gradual improvement occurred, and in 1870 Nevada County was one of the most flourishing mining counties in the state, and from that time forward this county has maintained its place among the leading gold producing counties of the state, regardless of the many depressing causes with which the industry has had to contend.

During the numerous periods of depression suffered by the quartz mining business, the attention and energy of the goldseekers were directed to the development of drift and hydraulic mining. As has been previously stated, hydraulic mining had its inception at American Hill, Nevada County. Many of the great improvements in the equipment of the hydraulic mines originated in the mines of Nevada County. From 1863, when the hydraulic Giant was perfected, this method of mining became increasingly important, reaching its maximum output with the completion of the costly works of the mines of the San Juan-Bloomfield ridge in the late '70's and early '80's.

The North Bloomfield mine was washing a gravel deposit 400' in height, 600' in width, and had worked 5000' of this great channel when closed by injunction. From 1863 to 1880 a number of the companies operating in this county had invested from $1,000,000 to $3,000,000 each in tunnels, reservoirs, ditches, flumes and other improvements. With the suppression of hydraulic mining, Nevada County, in which were the largest properties and producers in the state, was dealt the severest blow ever inflicted upon her mining industries, and a period of general depression ensued.

Some of the largest properties tried to continue operations under the permits and restrictions of the Debris Commission as provided by the Caminetti Act (1893), but this solution of the problem failed to secure the relief for the hydraulic mines that was anticipated. The mines that had reopened finally closed one by one until at the present time very few hydraulic mines are being operated in Nevada County, and their output is negligible compared with the total gold production. If, as previously outlined, hydraulic mining can again be resumed, Nevada County will produce millions of dollars from this source, which, added to the steady annual production of the quartz mines, will keep this county for many years the banner gold-producing county of the state.

# MINING DISTRICTS.

## FRENCH CORRAL MINING DISTRICT

### Introduction.

The French Corral Mining District is located in the northwestern part of Nevada County, ten miles northwest of Nevada City, the terminal point of the Nevada County Narrow Gauge Railroad. It includes the region about French Corral, Bridgeport, Allison Ranch, Rays Ranch, Montezuma Hill and Purdon Bridge.

A great deal of placer gold has been produced from the Tertiary river gravels within this district. The Neocene channel of the Yuba River runs in a southerly direction from North San Juan through the district. The deposits have been mined extensively and are nearly exhausted, excepting at French Corral. The production at present is small compared with that of former years.

French Corral lies at an elevation of 1700 feet. The climate is mild with moderate precipitations of rain and snow. Parts of the district, not over 2000 feet in elevation, are as a rule densely wooded with digger pine and oak; above 2000 feet yellow pine, spruce, and fir predominate. The principal timber tracts are in the Oregon Hills, several miles to the north.

There is plenty of water in the district. The South Fork of Yuba River and its tributaries flow through the district, and join the Middle Fork of Yuba River west of Bridgeport. San Juan, to the north, is furnished with water from the ditch of the Eureka Lake Company, which carries water from the headwaters of Middle and South Yuba rivers.

### History of mining.

According to an estimate of the army engineers in 1891, 32,500,000 cubic yards of gravel had been excavated from the French Corral district, while 10,000,000 cubic yards remained. Almost continuous deposits from North San Juan to French Corral have been worked throughout, and large parts of these areas are now exhausted.

The total production of placer gold from Camptonville, French Corral, Nevada City and Grass Valley together, by drifting and sluicing, was $48,000 in 1908. In 1909 the gravel mines at French Corral, Grass Valley, Rough and Ready, and Smartsville, produced only $8,000.

**Bibliography.**

Lindgren, W., Tertiary Gravels of the Sierra Nevada, Prof. Paper No. 73, U. S. Geol. Surv., 1911, pp. 121–125.  U. S. Min. Res. 1905, p. 179; 1906, p. 191; 1907, pt. 1, p. 215; 1908, pt. 1, p. 341. U. S. Geol. Surv. Folio No. 18, 1895.

**Topography.**

The region, in general, is one of low slopes.  The cañon of the South Fork of Yuba River, however, near French Corral, has a slope of 1000 feet in one-half mile; the cañon of the Middle Fork of Yuba River, forming the northwestern boundary of the district, is also precipitous but not so deep.  The area between the forks of the Yuba River may generally be considered a plateau cut by late erosion; the surface is broken by numerous hills which rise several hundred feet above the general level.  The Middle and South forks of the Yuba meet at a point about 3 miles west-southwest from French Corral; just east of this point they are separated by a hill 400 feet in elevation.  Northeastward the ground rises on a gradient of about 1000 feet to a mile. Many low hills and mounds break the general slope.  The highest of these is Montezuma hill, 2800 feet in elevation, overlooking the South Fork of Yuba River to the south.

**Geology.**

The rocks occurring in this district are: granodiorite, gabbro-diorite, diabase, augite-porphyrite, amphibolite schist, serpentine, pyroxenite, rhyolite and andesite.

Granodiorite is exposed over the greater portion of the district.  A belt of diabase and porphyrite changing to amphibolite schist, and another of serpentine, extend in a northerly direction through the district; gabbro-diorite is exposed to the east of these.

The southern end of a body of diabase and porphyrite occurs at French Corral, from which the belt continues northward through Oregon Hills; within the district the width is not greater than one mile.  To the west this belt is in contact with granodiorite of which the southwestern portion of the district is composed.  A porphyrite-diabase belt extending through the district from Newton is in contact with granodiorite on the west and is separated from the French Corral diabase belt by a wide tongue of the same rock.

Gabbro-diorite occurs to the east of the diabase area but they are in contact only at Owl Creek, for in other places a long, narrow body of serpentine intervenes.  The southern end of the gabbro-diorite belt is over a mile wide; it pinches out to the north where the granodiorite comes in contact with diabase and amphibolite schist. The grano-diorite, gabbro-diorite and diabase belt extend downward from the surface, while the andesite-breccia on Montezuma Hill and on Bunker Hill is a surface capping. A short distance above Bridgeport a dark diorite of the gabbro-diorite formation is in sharp contact with grano-diorite, and appears to have been brecciated by the latter. At other places, the same area of diorite grades into granodiorite.

Amphibolite schist, sometimes called greenstone schist, occurs in the eastern part of the diabase belt. It is an altered igneous rock, derived from augite-porphyrite by the changing of augite to secondary dark green hornblende. The change is thought to have been brought about by dynamic-metamorphism along a zone of shearing which extends through the district. The action probably occurred during the post-Jurassic period of mountain building. The western part of the diabase belt is composed of augite-porphyrite.

In the southeastern part of the district two bodies of pyroxenite occur in gabbro-diorite. One lies parallel to and on the north side of the eastern branch of Rush Creek; the other occupies a circular area one-half mile south of that creek. The pyroxenite in these small areas is generally coarse grained, dark green in color, and grades into gabbro.

The Calaveras formation is represented only by a small area in the southern extremity of the district. The closely folded slates and quartzites are the oldest rocks present, being of Carboniferous age. The age of the larger portion of the igneous rocks is late Jurassic or early Cretaceous.

The granodiorite and gabbro-diorite may each be regarded as a geologic unit; probably having been intruded within a short time of each other. The granodiorite is relatively younger. The question of the relative ages of the granodiorite and of the diabase and augite-porphyrite groups has not been definitely determined. Dikes of diabase and augite-porphyrite intrude the granodiorite which extends through the central part of the Smartsville quadrangle.

During the Tertiary period the rocks which now form the bedrock series, and carrying the gold-bearing quartz veins, were extensively eroded, as the main channel of the Neocene Yuba River ran through this district. Auriferous gravels accumulated during the Neocene epoch. At the end of the Neocene, rhyolite flowed from the volcanoes to the east. The rhyolite consists of white or light colored sand, some-times packed to a soft stone, and sandy clays interstratified with gravels. Evidence of a rhyolite flow occurs on Montezuma Hill. Flows of ande-site followed that of the rhyolite; these flows once covered large areas. but only small amounts have been left on hill tops and ridges, subsequent erosion having worn away the greater part.

### Mineral deposits.

Gold is found, within this district, in three forms: (1) As placer gold in the Tertiary river gravels; (2) In quartz veins; (3) As impregnations of auriferous pyrite in granodiorite and greenstone.

In the past the largest production of gold was from placer gravels, which were mined very extensively in this district. The Neocene channel between North San Jaun and French Corral is plainly marked by a succession of gravel areas. Along the course of the Neocene channel the bedrock rises on both sides, showing a comparatively narrow and steep river valley. From French Corral the channel probably extended south to Smartsville, where it turns to the east. Between these two places nearly all traces of gravel have been removed by the present river.

The gravels at French Corral deposited previous to the period of volcanic activity consist of well-rounded pebbles of quartz and siliceous metamorphic rocks, with some sand. In the banks they exhibit white or yellowish color. The thickness of the gravel between French Corral and San Juan averages 150 feet, the maximum being about 250 feet. The deposits continue northeastward for two miles toward North San Juan, and have a bedrock principally of granodiorite. A quarter mile from the end of this deposit another extends in the same direction for about a mile, and rests on bedrock composed chiefly of amphibolite schist. Neither of these deposits is covered with andesite.

A gravel deposit mined on the north side of Montezuma Hill has a bedrock of granodiorite. The gravels are covered by heavy bedded clays; these are covered with rhyolitic sand, which is, in turn, covered by andesite breccia. Drifts have been run from northeast to southwest under the volcanic masses. The deposit is probably a preserved portion of the Harmony channel, which drained northwestward from the region east of Nevada City, and joined the old Yuba River.

For a distance of a mile and a half east-southeast from Purdon Bridge a deposit of auriferous gravel, with bedrock of granodiorite has been mined. To the east the rocks form a hilltop composed of andesite breccia, which covers part of the gravels.

Gold quartz veins occur in the southeastern part of the district, about four miles southeast of French Corral. They occur not far from the contact of gabbro and pyroxenite, strike northwest, and dip northeast. In general, the veins of the region are quartz fillings in fissures, the quartz containing native gold and metallic auriferous minerals; the gold is usually concentrated in ore shoots.

Southeast of French Corral are two zones of rock containing considerable gold as an impregnation. One of these zones begins a half mile southeast of French Corral in granodiorite, curving and narrowing in the same direction for about two miles and extends into amphibolite schist. This zone is over a quarter of a mile wide at the north end. A second zone is narrower, extending for about two miles in a nearly north-south direction in amphibolite schist. A small but similar zone occurs one-half mile east of Birchville.

These zones are thoroughly impregnated with auriferous iron pyrites, without containing any quartz veins. The impregnation was probably caused by hot mineral waters at one time permeating the rocks. In places where these zones have been subjected to weathering, the gold in the pyrite has been freed and concentrated, thus forming large bodies of low-grade, free milling ore. Such a deposit was worked in 1895 by the Boss mine at Sweetland, where impregnated amphibole schists occurred.

## GRANITEVILLE MINING DISTRICT.

### Introduction.

The Graniteville district lies in the north-central part of Nevada County, 31 miles east-northeast from Nevada City, a station on the Nevada County Narrow Gauge Railroad. The distance by stage road from Nevada City to Graniteville, by way of North Bloomfield and Moore's Flat, is about 32 miles. The elevation of Graniteville is 4900 ft. The climate is characterized by heavy winter rains and snows, and by dry warm summers. The region in general is timbered with yellow and sugar pine.

The region includes both veins and Tertiary river gravels. There is a continuous string of veins occurring in the granites south of Graniteville, striking approximately north and south. The gravel deposits are in the immediate vicinity of the village.

### Bibliography.

Lindgren, W., Tertiary Gravels of the Sierra Nevada, Prof. Paper No. 73. p. 141. Min. Res. of the U. S., 1905, p. 179; 1906, p. 191; 1907, pt. 1, p. 215; 1908, pt. 1, p. 341; 1909, pt. 1, p. 280. Topographic sheet of the Colfax quadrangle, U. S. Geol. Survey Folio No. 66, 1900.

### Topography.

The region includes part of the divide between the Middle Fork and the South Fork of Yuba River. To the east are many lakes and considerable morainal material left from the Pleistocene glaciation period.

Graniteville is on a rather broad flat ridge which trends in a northeast direction, separating Cañon Creek on the south from the numerous transverse tributaries of the Middle Fork on the north. The elevation of Graniteville above the two cañons is from 1500 to 2000′, both being steep and V-shaped. Two miles southeast of Graniteville a ridge of granite peaks reaches an elevation of 6600′.

### Geology.

The Blue Cañon slates are the oldest known rocks in the Graniteville region, they being of Carboniferous age. They were folded and eroded at the end of the Paleozoic and again at the end of the Jurassic period. Accompanying the latter period of folding a great intrusion of granite took place. During the Tertiary period various movements occurred, the country was eroded, and Neocene gravels accumulated. At the end of this period an extensive flow of andesite covered the region; this is probably of the same flow as that found in other parts of the region. It was later eroded, leaving a cap on some of the present high regions, and in some places covered the old Neocene auriferous gravels.

The western third of the region is composed of Blue Cañon slates, being a continuation of that formation from the Alleghany and American Hill districts. The eastern part is granite; the contact with the slate striking north and south through Graniteville to a point 4 miles to the south of that village, where it turns about 35° to the east. Granite also composes the remainder of the western area. A slender arm of glacial detritus covers the granite two miles west of Graniteville.

### Mineral deposits.

The mineral deposits are both lode and placer. Placer gravels occur in the immediate vicinity of Graniteville and also 2¾ miles west, at Shands. At this latter place the usual andesite covering has been completely eroded, and the gravel rests on a bedrock of Blue Cañon formation. The areas are comparatively small, not over 100′ thick, composed of well washed pebbles, and covered by sub-angular gravel. The deposits near Graniteville are also of sub-angular gravel, and north of town the gravel outcrops below andesite, but is thin and irregular. It is possible that a small Neocene river channel of steep slope extends for some distance under the andesite cap.

The veins are chiefly of the fissure type, and occur in granite and slate. The most productive vein, called the California, or Gaston, occurs at a granite-slate contact, 2 miles south of Graniteville.

The veins of this district are part of the same north-south belt which in the Colfax quadrangle contains the most productive veins of that region. In the Graniteville district the lodes as a whole are not nearly so valuable as in the Alleghany district, due to the absence of rich ore shoots. The quartz veins probably were formed during early Cretaceous time.

There are two important vein systems south of Graniteville. The group to the east is composed of a continuous series of outcrops and forms an approximately straight N.-S. line, which if continued, would pass one mile east of the town. These are fissure veins in granite, and none have been very productive. The veins of the western group do not appear to be so continuous, although all the outcrops are N.-S. This group occurs in the Blue Cañon formation, and extends southward to form the important Washington Group at the cañon of the South Fork of Yuba River. The veins strike N.-S. and dip from 50° to 70° E.

The ore, compared with that of the Alleghany district, is of low grade; there being no rich ore shoots. The gold is nearly all free milling, but some occurs in pyrite, and concentrated sulphides often run from $50 to $60 per ton.

The Spanish mine is four miles southwest of Graniteville. The lode is from 20′ to 30′ wide, and is composed of slate-quartz stringers. It is joined on the west by a narrow dike of amphibolite schist. The ore is low-grade, and averages from $1 to $2 per ton.

The contact veins to the south of Graniteville have been the most productive of the district. Among these are the Rocky Glen, occurring near the slate-granite contact; and the California, or Gaston, occurring at the contact. The main Gaston vein has been followed to a depth of 1800′, the dip being 50° E. A well-defined hanging wall consists of granite, while the foot-wall is of slate. The white quartz vein contains gold, galena, and iron and copper sulphides. At one point on the vein, a 10′ foot-wall vein follows the contact while a 4′ hanging wall vein strikes into the granite. The granite separating the two veins. is considerably shattered, and has been enriched by the replacement of the feldspar, hornblende and biotite, by gold-bearing quartz and sulphides. This enriched granite is called 'gray' or 'blue' quartz by the miners, and is worth about $5 per ton, while the vein proper carries $10 in gold per ton.

Three miles west of Graniteville is a gold-quartz vein outcropping for about 1300′ and crossing the Middle Fork of the Yuba River. It occurs in slate of the Blue Cañon series, near the serpentine contact. The quartz carries considerable stibnite, as 6″ stringers, similar to that in the Plumbago mine in the Alleghany district. Parts of the serpentine wall rock are altered to mariposite and ankerite.

Two miles west of Graniteville there is a fissure vein in clay slate traceable for over a mile along the outcrop. The vein dips 75° E. and is from 1′ to 9′ in width; the sulphide content is less than 1%.

## GRASS VALLEY MINING DISTRICT.

**Location.**

The most important and prolific gold producing district in Nevada County, and in fact in the state, is that of Grass Valley, situated in the heart of the county at an elevation of 2400 feet. Grass Valley is a station on the Nevada County Narrow Gauge Railroad about 16 miles northwest of Colfax, at which point all trains connect with the Southern Pacific Railroad for San Francisco and eastern points. There are a number of trains daily from San Francisco to Grass Valley and the trip is made in about 7 hours.

The mines lie within a radius of 4 to 5 miles of the town and are easily accessible by excellent roads at all seasons of the year.

The climate in this district is ideal. The summers are tempered by cooling breezes and snow in winter is the exception rather than the rule, remaining on the ground for only a few days at the most. A wide variety of horticultural and agricultural products are raised abundantly in the vicinity.

Heretofore water power has been used at most of the mines, but in recent years this form of power has been gradually superseded by the more economical and dependable electricity, which is developed in the upper reaches of the rivers from waterpower utilized under more efficient conditions.

Timber for mining purposes is obtained at reasonable cost from the lumbering districts of the county.

**History of the district.**

In order that the description of each mining district may be complete in itself, a brief resumé of the history of mining in this district will be given herewith.

Within a comparatively short time following the discovery of gold at Coloma, Eldorado County, on January 19, 1848. placer gold was found in Wolf Creek and other streams in the vicinity of Grass Valley. In 1849 and 1850 many miners were working the extremely rich but comparatively limited shallow placer deposits of the district. The rapid exhaustion of the placer 'diggins' turned the attention of the miners to the gold-bearing quartz veins, to which source the gold had been traced in the placer mining operations. The exceptional richness of the oxidized ore in these veins caused lode mining to be more rapidly developed in Grass Valley than in any other district in California. Gold bearing quartz was first found in the fall of 1850 on Gold Hill, from the top of which it is reported over $2,000,000 was taken from within a few feet of the surface. Indeed the gold was so plentiful that the miners were afraid that gold would soon be demonetized and the lode claims were limited to blocks 100 feet square. The discovery on

Gold Hill resulted in the rapid development of other quartz veins on Massachusetts Hill, Ophir Hill, New York Hill, Granite Hill, and Union Hill. The richness of these discoveries led to intense excitement, and in the following few years, most of the quartz veins which have since become famous producers were located. The Empire mine, which has the proud record of having worked without cessation from its discovery to the present day, was located in 1850 and in the following year the North Star and Eureka mines were discovered. In 1851 the first crude stamp mill was erected in Boston Ravine.

As in the other districts of California, owing to the lack of knowledge regarding geoglogical conditions and metallurgical treatment, most of the quartz mining ventures in Grass Valley and Nevada City proved disastrous failures and this form of mining suffered a momentary setback. The success of a few of the richer and better managed lode mines gradually caused a revival, but the Comstock excitement, with the resultant exodus of men and capital, and the flooding of the Grass Valley mines during the severe winter of 1861-'62, caused another relapse. Again the industry recoverd from this period of depression and in 1867, according to J. Ross Browne (Mineral Resources West of the Rocky Mountains), 248 stamps were dropping in the Grass Valley district, which crushed 71,420 tons of ore yielding an average of from $30 to $35 per ton in gold.

In the year 1869 the Eureka, Idaho, Empire, North Star, and a number of the smaller mines produced $1,696,203. From this time forward, the quartz mining industry made rapid strides in the methods of mining and metallurgical treatment. The prosperity of the Grass Valley district continued until 1874 when the North Star mine was closed "because it was worked out." This was followed by the closure of the Eureka mine in 1877. In the meantime, to compensate for the curtailment of production caused by the closing of these mines, the Idaho mine had become one of the most prolific producers in California. Ore was first encountered in 1867, at a depth of 300 feet below the surface, and until 1894 this mine produced from $200,000 to $1,000,000 per year or a total of $11,873,553, paying during this period $5,017,-083.58 in dividends. In 1894 the Idaho passed to the Maryland company as a result of litigation, but did not long survive the change in ownership and management. The mine produced from 1894 to 1901, $1,246,020, but in this latter year it was closed. In 1904 it was reopened and worked superficially until 1914 when it was again closed.

The Empire mine, owing to exceptionally rich ore and good management, did not suffer the usual vicissitudes of fortune of the other early producers of the district, and produced from 1854 to 1863 inclusive, $1,056,234. After the installation of a new plant and equipment, it continued to produce at the rate of from $100,000 to $250,000 per year until 1878. Up to this time the mine had produced $2,967,315.85. At this period in the mine's history three well known mining engineers advised that it be abandoned as "worked out," but W. B. Bourn, Jr., decided that the mine warranted further developments; a new company was formed and extensive development work uncovered rich ore bodies. Later a 'barren zone' was encountered below the 1500 level and a 'split' on the main vein occurred below the 1700 level, owing to this fact operations from 1893 until 1898 were carried on at a loss. In 1898 Mr. Geo. Starr became manager and a campaign of energetic development rejuvenated the property, and since that time the Empire mine has been one of the best managed and greatest gold producing mines of the state. At the present time the Empire mine, including the production of its latest acquisition, the Pennsylvania, is turning out over $1,500,000 per year and making an estimated net profit of over $1,000,000 per year.

The North Star mine, after an idleness of 10 years, was reopened in 1884, by Mr. Bourn, who, having succeeded in bringing the Empire to a flourishing condition, had faith in the future of the Grass Valley mining district. From 1884 until 1894 this mine produced $2,500,000. In 1895 all the ore bodies which had been developed were worked out and the shaft and new development work proved to be in the usual barren zone. Under the efficient management of James D. Hague and A. D. Foote, however, other ore-bodies were opened, and the Central shaft was sunk encountering the North Star vein at a vertical depth of 1650 feet or at a point 4000 feet on the dip below the outcrop of the vein. Since that time the North Star has experienced an exceedingly prosperous period and the vein has been explored to a depth of 6300 feet on the dip. The North Star mine is producing in the neighborhood of $1,200,000 per year from ore which for the most part is being stoped above the 4000' level.

The histories of the North Star, Empire, and Idaho mines are typical of the other mines of this district, and of California mines in general. Since 1850 every producing quartz mine in this district, and for that matter in California, with the notable exception of the Empire mine, has at some period in its history been closed by reason of poor management, the exhaustion of their known ore-bodies, lack of sufficient development work or the encountering of the so-called barren zones which are bound to occur in all California mines. Even the Empire mine has, at different periods in its history, been condemned by competent mining engineers as worked out, and has only been saved from closure by the abounding faith of Mr. W. B. Bourn, in the possibilities of the mines of Grass Valley district in depth, and by the efficient management of

Mr. George W. Starr, for many years manager of the property. A perusal of the detailed reports on the Empire, Pennsylvania, North Star, Eureka, and Idaho-Maryland mines will give the reader a more detailed insight into the 'ups and downs' of the various properties in the district and will show the great debt that the Grass Valley and Nevada City districts owe to a few men who were far-sighted enough to see the possibilities of the veins of the district in lateral extension and in depth, and who were not discouraged by the exhaustion of an ore-body, or by a barren zone, but were willing to risk the heavy expenditures necessary to sink deep shafts and to accomplish the exceptional amount of 'dead work' necessary to prove the veins in depth. To such men as W. B. Bourn, James D. Hague, Geo. W. Starr and Arthur DeWint Foote, the Grass Valley Mining District owes a debt of lasting gratitude. It is a pleasure to record that the faith, foresight, and perseverance of these men have been amply rewarded by the past and present productions of the great Empire and North Star mines, which are today among the most famous gold mines of the United States and which from the recent deep developments, promise to continue their prosperity for many years to come.

### Production.

According to the carefully compiled figures of F. M. Miller, and later statistics quoted on a preceding page, Nevada County has produced in gold about $265,000,000 of which probably about $120,000,000 has been contributed by the mines of the Grass Valley District. The mines of the Nevada City-Banner Hill district have produced in addition $50,000,000 in gold.

PAST PRODUCTION OF THE GOLD-QUARTZ MINES OF THE GRASS VALLEY MINING DISTRICT FROM 1850 TO 1918 INCLUSIVE.

| | |
|---|---:|
| Eureka-Idaho-Maryland | $19,131,000 |
| Empire-Pennsylvania | 29,500,000 |
| North Star, and other consolidated mines (present company) | 21,000,000 |
| North Star, and other consolidated mines (previous operators) | 11,725,000 |
| Brunswick | 1,800,000 |
| Union Hill | 1,000,000 |
| Empire West | 3,500,000 |
| Allison Ranch | 2,500,000 |
| Sultana-Osborne Hill | 6,000,000 |
| Norambagua-Prudential | 2,250,000 |
| Golden Gate-Spring Hill | 2,000,000 |
| Other Properties | 8,000,000 |
| Total | $108,406,000 |

### Costs.

On account of the ideal mining and climatic conditions in this district, the working costs are exceptionally low for such narrow veins. Total mining, milling and overhead cost at the North Star mine was $5.25 per ton in 1913, and while the Empire company refused to divulge their production or costs, it is generally conceded that their total cost

3—46900

is about 50¢ per ton lower than those of the North Star. These costs have increased to over $7.00 a ton in 1918. For detailed data regarding costs see the description of the North Star and other mines.

### Bibliography.

California State Mining Bureau publications. U. S. G. S. publications. Lindgren, W., Gold Quartz Veins of Nevada City and Grass Valley districts: 17th Ann. Rept., pt. 2, 1896, pp. 1–262; Lindgren, W., Tertiary Gravels of the Sierra Nevada: P.P. 73, 1911; Mineral Resources, 1905, to date. Folios No. 18, Smartsville, and No. 29, Nevada City Special.

### Topography.

*Relief and Drainage.* The Grass Valley district lies at an elevation ranging between 2000 and 3000 feet. The topography of the district, considered as a whole, may be described as an undulating plateau, characterized by a number of low, flat and irregular ridges and hills culminating toward the east in the Ophir-Osborne Hill ridge, which attains an elevation of 3080 feet.

The drainage of this plateau is effected by Wolf Creek and its branches. In the northeastern part of the district the North and South forks of Wolf Creek, both of which have a westerly course, are separated by the Idaho-Maryland-Brunswick ridge, which rises from 200 to 400 feet above the bed of the streams. These branches unite within the town of Grass Valley to form Wolf Creek, which from the point of junction is turned abruptly to the south by a low irregular ridge. This westerly barrier extends from Alta Hill at the northwestern corner of the district to the Perrin mine, a total distance of three miles. Wolf Creek follows along the eastern base of this irregular ridge from 100 to 300 feet below its crest. To the east of Wolf Creek there is a gradual rise until the Osborne Hill divide is reached.

### STRUCTURAL AND HISTORICAL GEOLOGY.

The geology of the district, in common with that of the rest of the Sierra Nevada, has been discussed so elaborately in the publications cited, that it seems superfluous to repeat here what can be found in detail in the Colfax, Smartsville, and Nevada City folios, and the Seventeenth Annual Report of the United States Geological Survey.

Briefly stated, the rocks of the Sierra Nevada have been divided into two great series, called the bedrock series, and the later superjacent series. The former includes older sedimentary rocks, which have been tilted from an originally horizontal position till now nearly vertical, and have been greatly metamorphosed, together with older igneous rocks and metamorphic members derived from them. The granodiorite, so important at Grass Valley because of its occurrence in the deeper levels of the mines, is considered to be a later intrusion.

After the final erosion of the old surface of the bedrock series, the superjacent series of sedimentary rocks, volcanic flows and gold-bearing gravels were laid down nearly horizontally over great areas of the western Sierra Nevada. This series is of much interest in most of the county, but of minor importance in a discussion of the Grass Valley district. Only one area of Tertiary auriferous gravel has been mined to any extent here. This is the Alta Channel just north of town. There are a few other short segments of channel a half mile to a mile southeast of town which have been opened by inclined shafts and hydraulic pits. It seems appropriate to limit the discussion of geology in this district to a consideration of the veins, which have been so successfully exploited here. As one approaches Grass Valley from the western foothills, the first gold-bearing veins are encountered near Rough and Ready lying in a north and south belt of massive amphibolite from 1½ to 2 miles in width, which has probably been derived from diabase and the gabbro-diorite which flanks this area of amphibolite on both the east and west. Lying east of the gabbro-diorite is a long narrow strip of clay slate, and siliceous slates of the Calaveras formation. This area of sedimentary rocks is bounded on the east at the northern end by the diabase in which the famous North Star, Massachusetts Hill, Gold Hill, New York Hill, Rocky Bar and many other veins outcropped. To the east at the southern end of this belt of slates, lies an area of granodiorite. The east and west contact of this with the diabase occurs just south of the North Star vein. In this area of granodiorite occur the Omaha-Wisconsin-Hartery systems of west-dipping veins and also the Allison Ranch vein system, one of the greatest gold producers of the district in the '60's. Farther north the Pennsylvania-W. Y. O. D. and Golden Center veins are found dipping to the west. This granodiorite which has a maximum width of about 2 miles and a length of 5 miles extends from Forest Spring on the south to the serpentine, slate and andesite area lying just north of the city limits of Grass Valley.

East and south of this granodiorite belt, is found the great porphyrite-diabase area, generally dark green to greenish gray rocks, having an average width of 5 miles and extending in a southerly direction from Grass Valley to the Nevada-Placer county line. In this area occur the outcrops of the Ophir Hill- (Empire) Osborne Hill complicated vein system. A narrow belt of Calaveras slates, overlain for the most part by a blanket of andesite, separates this belt on the northeast from a similar diabase-porphyrite tract which forms the hanging wall of the famous Eureka-Idaho-Maryland vein, and which, altered to amphibolite schist by metamorphism, also forms the hanging wall of the Brunswick mine.

This porphyrite-diabase is terminated on the northwest by a surrounding mass of gabbro, while the derived amphibolite is terminated on the north by an area of serpentine which forms the footwall of the Eureka-Idaho-Maryland vein, and extends from this point in a northwesterly direction to Newton, where the dike-like mass is completely surrounded by gabbro. The Coe, Spring Hill, Golden Gate and Alpha system of north-dipping veins occurs in this serpentine associated with small, irregular dikes and masses of diabase and amphibolite schists. These mines have been worked to considerable depths with more or less success.

### Fissure Systems.

*Origin.* A careful study in detail of the fissures of the Grass Valley district leads to the conclusion that the fracturing and jointing was caused by successive compressive stresses exerted mainly from the east and west and from the north and south. These forces caused conjugate fractures in the homogeneous igneous rocks, with the same strike but of opposite dip.

The fissure systems of the Grass Valley district may, therefore, be divided roughly into four groups.

*Class 1.* Those veins having a northerly and southerly strike and a westerly dip. The Dromedary-Granite Hill-Omaha, the Pennsylvania, W. Y. O. D., Allison Ranch, and the Empire-Osborne Hill vein systems belong to this group.

*Class 2.* Those veins having a northerly and southerly strike but dipping to the east. These conjugate veins were formed by the same stresses and contemporaneously with the fissures in Class 1. The Bullion, the Gold Hill-Massachusetts Hill, the Great Eastern-Norambagua veins lying east and south of the Allison Ranch mine belong to this class.

*Class 3.* Those veins having an easterly and westerly strike and dipping to the south. The famous Eureka-Idaho-Maryland, the South Idaho-Brunswick-Union Hill veins are members of this class.

*Class 4.* Those veins having an easterly and westerly strike and dipping in a northerly direction. The North Star vein is an example of this group.

Classes 3 and 4 each have their sub-group of conjugate fissures but these are of less importance.

These fissures form a complex of branching and interlinked veins and the correlation and relative ages of the different groups is more or less problematical. A summation of the available data, however, leads to the following conclusions: The veins of Class 1, which include the Allison Ranch, Omaha, Pennsylvania, Empire veins, and Class 3, including the Eureka, Idaho, Maryland, Brunswick, Union Hill and Gold Point, are undoubtedly the main fissure systems. The conjugate veins of Class 2 are, of course, contemporaneous with those of Class 1, but as a rule are narrow and have not proven as large producers of gold as have the Class 1 veins.

According to Lindgren,[1] the Idaho-Maryland-Brunswick veins systems (Class 3) are probably of the same age as the north and south veins of Class 1, and the same authority gives the probable age of the North Star group (Class 4) as belonging to a later period. Recent developments in the deeper levels of the North Star, Pennsylvania and Empire mines, however, point to the conclusion that the North Star system should be correlated with the Empire-Pennsylvania group, thus making all four classes the result of the same geological epoch prior to the mineralization of veins.

From reliable information regarding recent developments, it is evident that the North Star shaft at a depth of 6300 feet on the northeasterly dipping vein has intersected the west-dipping Pennsylvania ledge. The 6000-foot east drift from the North Star shaft is also reported to have encountered this vein. As far as can be ascertained, the exploratory work undertaken by the North Star company has to date, failed to find a continuation of the North Star vein beyond the Pennsylvania vein, but it is said that the latter vein continues downward on its regular dip. The Pennsylvania vein has also been developed southward by the North Star company, who have also recently purchased the Bonivert ground of 100 acres lying south of the Pennsylvania-W. Y. O. D. claims belonging to the Empire company. The Empire company a year or so ago, acquired the property of the Golden Treasure Mining Company, which is bounded on the west by the Bonivert ground, on the west and north by the W. Y. O. D., on the east by the Cassidy-Nevada claim, and on the south by the property of the Bullion Consolidated; and within the past few months have purchased the Berriman holdings comprising 160 acres of agricultural land. This property is bounded on the north by the Bonivert property of the North Star company, on the east by the Bullion and on the west by a narrow strip of agricultural ground separating it from the Omaha claim of the

---

[1] Lindgren, W., Gold quartz veins of Nevada City and Grass Valley: U. S. Geol. Surv., Ann. Report XVII, Part II, 1896.

Empire West group. The Berriman property lies about one mile north of the Allison Ranch mine. The foregoing facts lead to the conclusion that the Pennsylvania-W. Y. O. D. fissure is one of the main fissures of the district, and that the Allison Ranch vein is the southerly continuation of this fissure system.

The veins lying in the hanging wall side of the Allison Ranch are the southerly continuation of the Omaha fissure system, which belongs to the west-dipping, north- and south-striking general group which include the Pennsylvania and the parallel Empire veins.

The general conclusion is that all the fissures of the Grass Valley district have been formed during the same geologic period by compressive stresses exerted from different directions.

The total displacement or upward movement of the hanging wall (reverse or overthrust faults) which has taken place along the main fissures is comparatively small, with the possible exception of the Eureka-Idaho fracture. Extensive sheeting of the granodiorite and other igneous rocks has occurred resulting in not only parallel veins, but in the formation of minor fractures between the main lines of faulting. Thus the compressive stresses have been relieved by this sheeting, and the movement has not been excessive along any one fissure or line of weakness. For this reason the veins of the Grass Valley district are narrow or small when compared with the veins along the Mother Lode. Little mechanical alteration of the rock wall has taken place and the fresh country rock lies very close to the vein with little 'gouge' separating them.

*Influence of wall rocks.* As a rule the fissures in the diabase-porphyrite seem to be less definitely confined to one plane than are the same fissures after passing into the granodiorite; in other words the veins have more branches and are also slightly larger than in the granodiorite. The contacts of the various rocks have not been lines of weakness and therefore have had little influence on the fissure systems. The notable exception to this is the Eureka-Idaho, where the fissure follows the contact of the serpentine and gabbro-diabase. Here the upward movement of the hanging wall has been comparatively large and the mechanical and hydro-thermal alteration has been more extensive than elsewhere in the district.

*Faults.* Post-mineral faulting and sheeting has taken place and most of the veins of the district show evidence of slight displacements or bending by these barren cross seams, which have a general strike of east-northeast and a steep dip. These seams now act as watercourses for the circulation of vadose or surface waters.

## Vein Systems.

The vein systems are the result of the filling of the previously described fissure systems by quartz, carrying gold and auriferous sulphides.

*Genesis.* The slipping of one irregular and undulating wall of the fissure on the other, resulted in open spaces forming in the fissures through which circulated hot solutions carrying silica, hydrogen sulphide, carbon dioxide, gold and other metallic minerals.

These thermal waters were in all probability surface waters which penetrated to great depth, were heated, and returned to the surface where their mineral content was deposited in the open spaces of the fissures.

The gangue minerals, the metals, and the sulphides of the present veins were undoubtedly dissolved by these waters from the deeper portion of the granitic mass; for although these veins traverse different rocks, the mineral content of the vein is practically the same, proving conclusively that the minerals were not derived from the wall rocks of the vein. As the solutions approached the surface zone, their mineral content was deposited by reason of the unbalancing of the chemical compounds; by the absorption of certain elements through alteration of the wall rocks; by the mingling of chemically different waters in the different intersecting channels, and by varying conditions of temperature and pressure.

The effect of the wall rocks and the intermingling of the solutions were probably of the greatest importance in causing the deposition of the quartz, gold and metallic sulphides.

It must here be remembered that the portions of the veins now being worked were, at the time of their formation, probably a few thousand feet below the ancient surface of the country.

*Alteration of wall rocks.* As has been previously stated, some of the rocks of the district have been altered over large areas by dynamic metamorphism. The schists in the vicinity of the Brunswick mine which were derived from the normal diabase porphyrite are an example of dynamic-metamorphism; hydro-metamorphism is shown in the formation of the large areas of serpentine of the Idaho-Maryland area. Further, in the granodiorite and other igneous rocks extensive sheeting and jointing has taken place.

The fissures themselves were probably formed by sudden breaks in the rock and not by long continued movement along the same line of weakness as in the veins of the Mother Lode. The result has been the formation of a system of main fractures with branching and minor fractures linking together the main breaks. We therefore find along the breaks, breccias (broken and crushed fragments of the wall rocks) which have since been chemically altered by the vein solution and finally cemented together by the vein-filling quartz.

The chemical alteration of the wall rocks next to the quartz varies in intensity to a marked degree, the greatest amount of change having taken place in the case of the Idaho-Maryland; a smaller amount in the case of the Empire and North Star veins and least in the case of the Omaha and Allison Ranch veins where comparatively fresh grano-diorite is found close to the quartz vein.

*Ore deposits.* The gold bearing ores of the Grass Valley district are practically confined to the quartz veins and the altered wall rocks carry very little gold.

*Outcrops.* In general the outcrops of the veins in this district are inconspicuous and as a rule they can not be traced for any great distance on the surface. For example the Eureka-Idaho-Maryland vein has been worked for a distance of 6000 feet along the vein without a break, yielding about $19,000,000 and yet this wonderful ore-body outcrops for but a few hundred feet on the Eureka claim. This outcrop of quartz was very low grade and pay-ore was not encountered until a depth of over 150 feet had been reached. The vein can not be traced on the surface east of the Idaho shaft, although underground some of the richest ore has been taken from the pay-shoot for a distance of 3000 feet east of the shaft.

The famous Empire, Pennsylvania and North Star veins can only be found at intervals on the surface. The reason for this is that the veins are easily decomposed forming a soft reddish mass of limonite and quartz extending to an average depth of 150 feet below the surface. The sequence of the deterioration seems to be first oxidation of the sulphides with the consequent liberation of the associated gold, followed by a general loosening of the texture of the vein filling. As a rule the disintegration of the vein is slightly more rapid than that of the surrounding country rock; and, as the veins are small, the decomposition of the country rock covers the vein, making it exceedingly difficult to trace the lodes on the surface. This necessitated considerable exploratory surface work to locate them definitely.

*Form and structure of veins.* The main veins of the district are more or less regular in strike but local undulations occur and a vein may branch and these branches may reunite or one may gradually die out. On the dip of the vein the same branching may occur, or where there are two parallel veins a short distance apart, a spur may go off from one vein into either the foot or hanging wall, and join the other vein. Thus a system of branching and linked veins is formed and unless the greatest care is exercised in following the various branches, the main vein is likely to be lost. On their downward course at different points, the Empire, Pennsylvania and North Star veins all branch and a considerable amount of exploratory work was necessary before it was determined which was the main vein. In the North Star mine the drifts were driven following the vein and when a 'split' was encountered, as a rule, the largest and most promising branch would be followed. It was often found, however, that this branch would die out within a short distance and that the less-likely looking stringer would lead ultimately to an orebody. For this reason it is most important that extensive exploratory work be continually carried on, and it is especially advisable that cross-cutting be undertaken at frequent intervals, in order to be sure that the development work is being done on the main vein. In the North Star mine, the vein split on its downward course and the lower vein was followed and some rich ore was stoped therefrom. Subsequently a cave-in of the hanging wall revealed the fact that the hanging wall stringer which had been practically overlooked, had become a large and very rich vein. The stoping of this vein proved that a double ore-shoot had been formed with a horse of country rock between; and it was further found that the two branches reunited at a point about 400 feet below their point of division.

The walls of the vein are as a rule undulating curved surfaces, and the quartz is separated from them by a few inches of gouge or finely comminuted material derived by minor movement along the fissure subsequent to the formation of the quartz vein.

The veins vary greatly in width; the distance between the walls may vary from a mere seam to a maximum of 10 feet. Between the walls the vein filling is seldom composed of solid quartz. It is generally a mixture of brecciated, altered wall rock and calcite and quartz stringers with a vein of quartz from 1 foot to 3 feet in width usually found on the hanging or footwall, but sometimes on both.

The average width of the North Star quartz vein is about 2' to 2½' in the area stoped; in the Empire, the vein averages only about 18" in width. The maximum width of workable ore attained in these mines is from 8' to 10'. These swells in the orebody generally occur at the intersection of two veins or where two branches of the vein follow a nearly parallel course, separated by a few feet of altered or shattered wall rock.

Underground the veins have been followed by drifts in the different mines for distances varying from 1000 feet to a maximum distance of 6000 feet in the case of the Eureka-Idaho-Maryland mines. In each mine, however, the vein would, at various levels, close down to a mere seam for several hundred feet to finally open out again into a pay shoot.

In the North Star mine the 1900 level followed such a seam, which sometimes was so indistinct that it was followed only with the greatest difficulty by keeping the general course of the vein. At a distance of about 1000 feet this seam intersected other seams and developed a large and rich ore-body several hundred feet in length along the strike of the vein, and which was mined from the 2300 to the 1100 level. Similar conditions are found in all the veins of the district.

*Depth.* The various mines of this district have at the present time (January, 1918) attained the following depths on their respective veins:

The North Star mine was worked to a depth of 2400 feet by an inclined shaft on the vein; a vertical shaft 1650' in depth was then sunk which encountered the vein at a point 4000 feet on the dip of the vein; this vertical shaft was connected with the old incline by means of a 1600' arise. The vertical shaft was then turned and continued down on the vein until in July, 1915, it had reached a depth of 2300 feet below the 4000 level; thus developing the vein to a total inclined depth of 6300 feet below the outcrop.

The Empire shaft has reached a depth of 5000 feet on the incline and the vein is being developed by drifts at that depth. The average dip of the vein above the 3400 level is about 30° W. but below this level the vein rapidly steepens to a maximum of 55° W.

The Pennsylvania vein which has an average dip of 40° W. has been developed to a depth of only 2600 feet by the Pennsylvania inclined shaft, but has recently been opened by a cross-cut from the 4600 level of the Empire shaft, giving a depth of 3500 feet on this vein or a vertical depth of 1900 feet. The drifts from the lower levels of the North Star mine have also encountered the Pennsylvania vein at a vertical depth of about 2600 feet.

The Idaho-Maryland vein, which has an average dip of about 70° S., has been worked to a depth of 2000 feet.

In none of the foregoing mines has there been any appreciable diminution in the tenor of the ore, at the great depth reached; in fact the Pennsylvania ledge where it has recently been developed in both the Empire and North Star workings is as rich as in any of the upper levels. The Empire has also developed large bodies of high grade ore in its lower levels. Therefore there seems to be little doubt that the ore shoots in the Grass Valley district will continue in depth to the limit of economical handling.

*Relation to country rock.* As has been previously stated, both the Empire and North Star veins pass from the diabase-porphyry into the underlying granodiorite, the Empire at a depth of 1700 feet and the North Star at a depth of 3700' to 4000' measuring along the dip of the vein. In both of these mines this change in the character of the wall-rocks seems to have little effect upon the tenor of the orebodies or upon their size, with the possible exception that in the latter characteristic they seem in the granodiorite, to be more regular and with fewer branches.

### Ore Shoots.

*Form.* The ore shoots of the Grass Valley district are most irregular in form. Some of the pay shoots have been found at the surface, while the tops of other shoots have been encountered at varying depths. Of the latter class the Eureka-Idaho ore shoot is a typical example. This has been worked for a length of over 5500' along the vein and yet it does not come to the surface at any point. In the Eureka mine the pay shoot was not opened until a depth of 100 feet had been reached, while in the Idaho mine the shaft was sunk 300 feet before encountering the orebody.

The gold may occur in 'bunches,' 'patches,' 'chimneys,' or in large areas of irregular outline on the plane of the vein. Some of the veins show a marked tendency for the coarse gold to form in small bunches or pockets within the pay shoots with more or less barren areas of quartz between; the Gold Hill and Allison Ranch are examples. In other mines the ore shoots may be continuous along the vein for lengths varying from a few hundred to 2000 feet or more, with a like variation in depth on the dip of the vein.

*Surface enrichment.* The upper part or oxidized zone of these orebodies within a few hundred feet of the surface or above the water level, has in the past, shown considerable secondary enrichment due to mechanical and chemical concentration of the gold following the weathering and erosion of the veins. Below this surface zone, however, while there may be great variations in the tenor of the ore within the same ore shoot and also at different horizons, there can not be said to be any general diminution in value of the ore as depth has been attained.

*Barren zones.* The so-called 'barren zones' have been encountered in every mine at varying depths. Owing to the irregularity and lack of continuity in the pay shoots, again excepting the Idaho-Maryland, levels may be driven a thousand feet or more along a fissure before cutting an ore shoot; in the same way in sinking on the vein a barren area may be developed between the various pay shoots that generally occur on the plane of the vein in the same mine. In the Empire mine a barren zone was passed through from the 1500 to 2100 level; in this zone according to the statements of Mr. Geo. W. Starr, managing director, little ore was found and the mine was supposed to be worked out. The consummate faith of the owners, however, in the future of the Grass Valley district, caused the rejuvenation of the property by a campaign of systematic development and improvement.

The North Star mine and the Pennsylvania also had barren zones at various periods. The following is quoted from a report by Mr. Starr on the Empire West or 'Omaha' property:

"The Omaha Consolidated and the Empire West Mines under my management have, during the past seven years, been carrying on development in a zone of ground between the 1000 and 1400 levels and over a distance of 3000 feet. Over 10,500 feet of drifts, shafts and raises have been run and there is now in sight 140,000 tons of low grade ore developed. My reason for confining the development to this zone was that I thought that we would find a continuation of the rich orebodies that were known to have existed at the workings of the Omaha, Homeward Bound, Illinois and Wisconsin mines, and my plan was, after this ore was developed, to then erect a suitable plant and sink the main shaft.

The results of the above work now prove that we have developed a poor zone of ground and are now in the position of the Empire mine, when that company for eight or nine years labored through a poor zone between the 1500 and 2100 levels. I may also refer to the position of the North Star mines that had a similar zone of poor ground and there are a number of other properties in this district that have had the same experience. I firmly believe that the Empire West group of mines is characteristic of the mines in the Grass Valley district and will in depth prove valuable properties."

The secret of successful mining in the Grass Valley district, and in fact in California in general, is that when an orebody has been developed, instead of immediately paying out the net profits in the form of dividends, a reserve fund be created with which to do the exploratory work necessary to develop other pay shoots. The extensive development work done in the mines of this district has conclusively proven that, when one orebody has been found in a strong well-defined fissure, systematic and persistent development work will uncover other pay shoots, although it may be necessary to drive drifts from 500′ to 2000′ along the fissure following barren 'stringers,' or it may be necessary to sink hundreds of feet through a barren zone. Persistent exploratory work, however, with but few exceptions, has resulted in the blocking out of numerous pay shoots which may be connected by 'narrow necks' or may be entirely independent of one another.

Moreover in the cases where a complex vein structure has been formed and considerable sheeting has taken place, it sometimes occurs that parallel veins will be found either in the footwall or hanging wall of the 'main' vein; for this reason in such cases frequent crosscutting is advisable.

*Tenor of the ore.* In the surface or oxidized zone above water level, the orebodies due to secondary enrichment varied in value from $25 to $300 per ton. Below this zone the ore varies in value from a few dollars to $30, with occasional small bodies of very rich 'specimen' ore running thousands of dollars to the ton. The average value of the North Star ore for 1913 was $11 per ton milled and while the Empire refuses to divulge their production or working costs, it is likely that the average value per ton milled from the Empire and Pennsylvania mines is slightly greater than that of the North Star mine, probably between $13 and $15 per ton. In the case of the Empire about 40% of the rock crushed is 'waste,' or wall rock, carrying only a small amount of auriferous sulphides and gold.

### Ores.

The veins of the Grass Valley district differ from the other veins of the Sierra Nevada, and especially the Mother Lode veins, in that they are narrow but produce a higher grade of ore. The vein filling is composed for the most part of quartz, calcite and altered wall rock.

*Gangue minerals.* Quartz is the most important of the gangue minerals and as a rule is massive or shows partly formed crystals with only occasionally 'vugs' of perfect quartz crystals. The quartz is usually milky-white in color with a glassy luster, and contains many minute fluid-filled cavities of irregular outline, which were probably formed contemporaneously with the deposition of the quartz.

The ore in some cases has the typical ribbon structure due to the reopening of the fissure or to the crushing of the quartz by post-mineral movement along the plane of the vein. In other cases a banded structure of the quartz is found due to deposition of sulphides or new layers of quartz at various intervals, during the forming of the veins.

*Calcite.* Calcite occurs as an alteration product of the wall rock near the vein and also of the entrapped breccia of country rock between the vein walls. It is also found mixed with the vein quartz as small flakes and bunches.

*Scheelite.* This ore of tungsten is found in the Union Hill mine in paying quantities, where a 6″ vein of quartz and scheelite has been developed, with free gold occurring in the scheelite. It also occurs in small amounts in the ores of the Empire and Pennsylvania mines.

*Mariposite.* This vivid green, chromium mica occurs only in the Idaho-Maryland as an alteration product of the serpentine.

### Metallic Minerals.

*Sulphides.* As a rule the sulphide constituent of the ore averages between 2% and 4% of the total; it may be disseminated irregularly throughout the massive quartz or may occur in ribbons or bands generally parallel to the walls of the vein.

*Pyrite* is the most abundant sulphide in the ores, predominating to the extent of 80% of the total sulphide content.

*Galena* is a normal constituent of the sulphides and in a finely disseminated form generally accompanies the rich 'specimen' ore. Galena also occurs in cubical form but does not indicate rich ore. As a rule the amount of galena is small compared with the pyrite, averaging only about 3% to 5% of the concentrate.

*Chalcopyrite* occurs in about the same quantity as galena and is found in most of the quartz veins.

*Sphalerite* (zinc blende) also occurs sparingly in the veins of the district and is associated with rich ore. The miners refer to it as the 'mother of gold.'

*Arsenopyrite* occurs in a few veins in the district.

*Tellurides* of gold have also been reported in the concentrates from some of the mines.

*Gold* occurs associated not only with the sulphides but also in a free state in minute particles distributed throughout the quartz. It also occurs as leaves and coarse bunches at irregular intervals in the pay shoot. This rich ore is known locally as 'specimen rock.'

## LOWELL HILL MINING DISTRICT.
### Including Lowell Hill, Remington Hill, and Democrat.

This district is among the larger producers of placer gold, in that portion of Nevada County within the Colfax quadrangle. There are no gold-quartz veins known within the district.

Lowell Hill is in the south central part of Nevada County, four and a half miles in a straight line, and six miles by road northeast of Dutch Flat, a station on the Southern Pacific Railroad. A fair road to Lowell Hill, but indirect, runs first southwest to Little York, crosses the Bear River Cañon at a moderate grade, and follows northeast to Lowell Hill along the ridge northwest of the Bear River.

The climate of Lowell Hill and the surrounding region is warm and dry in the summer, with rainfall and some snow during the winter months. This district is within the Sierra Nevada forest zone, and vegetation consists chiefly of yellow and sugar pine, with spruce, fir, and oak on the lower slopes. At high elevations feed keeps green after it has dried in the lower valleys, and parts of the Colfax quadrangle are used as a summer range for cattle. Water is supplied by the Pacific Gas and Electric Co., which utilizes the headwaters of the South Fork Yuba River and lakes in the Truckee quadrangle to the east. There is much water in the various streams and in Bear River. Electricity may be used for power at a low rate per kilowatt-hour. Timber may be obtained from uncut portions of forests in the surrounding region.

Gold mining is the most important industry. Timber-cutting and horticulture are also of importance. Apple trees grow well at the elevation of Lowell Hill, 3900 feet.

**History of mining.**

At Liberty Hill, near Bear River, southwest of Lowell Hill, it was estimated in the year 1900 that 2,000,000 cubic yards of Tertiary auriferous gravels had been washed away while 1,000,000 cubic yards remained. At Lowell Hill hydraulic work is difficult, since heavy masses of clay cover the gravels. Considerable work has been done at the Planet mine, located south of Nigger Jack Hill, about 1½ miles east of Lowell Hill. Drift mining has been carried on at the Valentine mine and at the Swamp Angel mine, located opposite the Planet property.

At Remington Hill, located one mile north of Lowell Hill, it is reported that 1,750,000 cubic yards of Tertiary gravels had been excavated up to 1900, leaving 600,000 cubic yards, capped by clay and tuff, remaining.

**Bibliography.**

Lindgren, W., Tertiary Gravels of the Sierra Nevada, Prof. Paper No. 73, U. S. Geol. Survey 1911, pp. 146–147. Min. Res. of the U. S. 1905, p. 179; 1906, p. 191; 1907, pt. 1, p. 215; 1908; pt. 1, p. 341. U. S. Geol. Survey Topographic sheet Colfax. U. S. Geol. Survey Folio No. 66, 1900.

**Topography.**

Lowell Hill, situated at an elevation of 3900 feet, lies on a broad flat upland area which extends northeastward parallel to Bear River on the southeast. This divides Bear River from the upper waters of Steep Hollow Creek to the northwest. Two peaks mark this area, Nigger Jack Hill rises to an elevation of 4700' one mile east of Lowell Hill, while another peak one mile south of Lowell Hill has an elevation of 4520'. Remington Hill is located one mile north of Lowell Hill midway up the slope toward the upland area. The region about Remington Hill and Lowell Hill is drained by Steep Hollow Creek, which flows southwest between the two villages.

**Geology.**

The greater portion of the broad flat area in the northern part of the district is covered by andesite breccia, which rests on eroded slates and quartzitic sandstones of Blue Cañon formation. In the southern part of the district gabbro and serpentine occur to the west and to the east, respectively. These extend across the river into Placer County. The contact lines between the metamorphosed sedimentary deposits and between the intrusions themselves trend in a north-south direction. At Lowell Hill the dip of schistosity is vertical, and the strike is northwesterly along contacts with intruded igneous rocks.

The Blue Cañon formation consists of shales and sandstones which have been closely folded and changed by pressure to black fissile slates and dark gray, fine-grained, quartzitic sandstones having a rough schistosity. It is the lowest of any of the Carboniferous formations in the region. At the end of the Juratrias period these sediments were folded and lifted up, during which time certain sediments found in other portions of the Colfax quadrangle were deposited. Also at the end of the Juratrias period igneous intrusions occurred. The veins furnishing the gold now found in gravels are believed to have been formed at the beginning of Cretaceous time.

During the Tertiary period of erosion auriferous gravels accumulated, especially during Neocene or late Tertiary time. At the end of this period an extensive flow of andesite covered the entire region, the greatest thicknesses being reached where it filled cañons of a drainage system not dissimilar to the present one. Much of this has been washed away in the following cañon cutting process, but it is still found extensively on the flat interstream areas. In many places the buried Neocene gravels have been exposed by this Pleistocene erosion. In early Pleistocene time, scattered eruptions of basalt occurred and a small pear-shaped area of lava occurs $2\frac{1}{2}$ miles east-north-east of Lowell Hill, resting on Blue Cañon metamorphic rock.

### Rocks.

Gabbro is a dark green rock, consisting of pyroxene and greenish gray feldspar, the latter having a flinty fracture and generally an altered appearance.

A great serpentine belt traverses the Colfax quadrangle. It is a metamorphosed igneous rock, derived from peridotite, being dark green in color, compact, and with a dull to waxy luster.

Basalt is a black, fine-grained, and sometimes vesicular rock, often containing small crystals of olivine.

### Mineral deposits.

The only mineral deposit of present economic interest is the gold derived from Tertiary river gravels. No gold quartz veins are known to exist within the district.

A branch of one of the main Neocene river channels is thought to have extended through Democrat, Remington Hill and Lowell Hill, crossing the Bear River $2\frac{1}{4}$ miles south-southeast of Lowell Hill and joining a larger channel at Dutch Flat. This larger channel continues through You Bet and Scott's Flat to North Columbia, where a juncture is made with the North Bloomfield channel.

Gravel deposits occur along the branch channel in the Lowell Hill district; also small areas occur at Democrat and at Remington Hill. The most extensive deposit, at Lowell Hill, continues southward.

Rising rapidly from Elenore Hill, a point between Bear River and Little Bear Creek in Placer County, the channel extends to Liberty Hill. There the gravel is about 60 feet deep and consists of a lower half of blue gravel with reddish quartz gravel above. The blue gravel is full of large boulders of gabbro and serpentine.

The gravel at Lowell Hill is covered by heavy masses of light colored clay. It is 30 feet deep and the coarse bottom material is covered with a finer quartzose gravel. A depression filled with gravels of a few acres extent occurs at Remington Hill, north of Lowell Hill. The material is capped by heavy masses of clay and is similar, in other respects, to that at Lowell Hill.

The bedrock at both Lowell Hill and Remington Hill consists of the Blue Cañon formation, while that at Democrat is serpentine.

A small gravel hill called Excelsior occurs east of Democrat at a point between two forks of Steep Hollow Creek. This is beyond doubt on the same channel which occurs at Democrat. The bedrock is serpentine and Blue Cañon formation. To the north and northwest of this place the bedrock rises rapidly. It is not certain whether the old channel is covered by andesite breccia to the northeast or whether it extends along the present course of Steep Hollow Creek.

## MEADOW LAKE MINING DISTRICT.

Gold is the only mineral produced in this district. It occurs in auriferous sulphides of fissure veins. The production of this district has not been large on account of the low grade of ores and difficult transportation.

Summit City is located on the west side of Meadow Lake 15 miles in a straight line, and 30 miles by road, northeast of Emigrant Gap, a main line station on the Southern Pacific Railroad. It has an altitude of 7300 feet. Heavy snowfall during the winter months renders the roads nearly impassable for the greater part of the year, but during the summer months the roads are good.

Timber is abundant, pine and fir predominating. The slopes are covered with manzanita brush. Glacial lakes and numerous small streams assure a plentiful supply of water to the entire district.

### History of mining.

Mining on a small scale has been carried on in this district. Frequent attempts, made to exploit the veins in the section between Summit City and Old Man Mountain generally have not been successful, owing to the low grade of ores, the absence of free gold, the climatic conditions

4—46900

and the difficulty of communication.  At Jackson Lake, north of English Mountain, a vein is exposed in diabase porphyrite which is similar in character to the veins of Summit City.  The decomposed portions contained considerable free gold, which was successfully exploited during 1895 and 1896.

### Bibliography.

Min. Res. of the U. S. 1907, pt. I, p. 215.  Truckee topographic sheet.  U. S. Geol. Survey Folio No. 39, 1897.  Colfax topographic sheet.  U. S. Geol. Survey Folio No. 66, 1900.

### Topography.

Meadow Lake lies in a shallow draw, on a ridge with a general northwesterly trend, near the Nevada-Sierra county line.  This ridge rises to an elevation of 8498 feet at English Mountain, the northeast and southwest slopes of which are very steep.  Jackson Lake lies in a draw along the continuation of this ridge being northwest of and about 1500 feet lower than the summit of English Mountain.  Trending southerly from Meadow Lake is a ridge, the high points of which are Old Man Mountain, 7800 feet, and Signal Peak, 7860 feet.  Fordyce Creek has cut a deep cañon about 2000 feet deep between Old Man Mountain and Signal Peak.  The slope from Old Man Mountain is steep, while that to the west from Signal Peak is gradual.  South of the ridge flows the South Fork of Yuba River with its Rattlesnake Creek branch flowing southwesterly and cutting a deep cañon in the southern extremity of the ridge.  About three miles northwest of Old Man Mountain is a ridge called Black Mountains; it is about 2½ miles long, and trends northeasterly and southwesterly with an elevation of about 8000 feet.  To the west of Black Mountains numerous lakes occur at elevations of from 7000 to 7500 feet; to the north and northwest, trending from Summit City northwesterly between Black Mountains and English Mountain, is a broad valley occupied by Cañon Creek and several lakes.

The southern portion of the district is drained by the South Fork of Yuba River with its tributaries Fordyce and Rattlesnake creeks.  The northern and western portions are drained by Cañon Creek, which rises in the center of the district, and flows northwest then turning and flowing southwest into the South Fork of Yuba River.

### Geology.

By far the greater portion of this area is made up of granodiorite.  In the southern part is an exposure, 1½ miles wide and 2 miles long, of Sailor Cañon formation on both sides of Rattlesnake Creek, and extending northerly from the Nevada-Placer county line.  A narrow dike of granodiorite forking near the end, extends into this on the west side for a distance of about three-fourths of a mile. Three small dikes of granodiorite occur in the northwest portion. The granodiorite is in contact with Blue Cañon formation on the west and with a strip of Sailor Cañon formation on the northwest; the latter trends northwesterly and is about 2 miles wide. Lying to the east of the Sailor Cañon formation is an area of diabase nearly two miles in diameter making up English Mountain. On the south and east this is in contact with granodiorite. On the northeast the granodiorite is capped by a relatively large mass of andesite, which extends down to Summit City and nearly surrounds Meadow Lake. On the east the granodiorite is in contact with diabase-porphyrite which trends north-south.

In the southeast portion of the area a flow of andesite, covering the region north of Rattlesnake Creek extends, as a narrow strip, in a westerly direction towards the northeast corner of the Sailor Cañon formation. An area of diorite averaging one mile wide extends from Grouse Ridge through Summit City northward to the andesite contact. Two small areas, one of Sailor Cañon formation and the other of diabase-porphyrite, making up Black Mountains, are surrounded by diorite. The Sailor Cañon formation occupies a circular area one mile in diameter, nearly surrounded by diorite. The area of diabase-porphyrite is elliptical, lying to the northeast-southwest, about $\frac{1}{2}$ by $1\frac{1}{2}$ miles in contact on the north with granodiorite, on the southwest with Sailor Cañon formation, and otherwise diorite. Lying about $1\frac{1}{2}$ miles to the north of the last mentioned area of Sailor Cañon formation is a small lenticular shaped body of andesite capping the granodiorite. About $1\frac{1}{2}$ miles northwest of Old Man Mountain is a small mass of andesite capping granodiorite. Several exposures of morainal and glacial drift occur; one lies just south of the Sailor Cañon formation on Black Mountains, and another underlies the andesite. Other areas occur near Summit City north of French Lake, and between Fordyce Lake and Meadow Lake.

After the deposition of the Calaveras formation and before the deposition of the Juratrias, the Paleozoic series was closely folded and compressed. These latter exhibit a schistosity having a general north-northwest direction and a steep easterly dip. The Sailor Cañon formation was without doubt deposited unconformably on the upturned Calaveras formation. A mountain-building disturbance followed the deposition of the Juratrias, and the later beds were folded against the Carboniferous land masses and considerably compressed. During the latest Juratrias or earliest Cretaceous great intrusions of igneous rocks occurred. Some of them, as for instance the diabase of English Mountain, was probably intruded during the deposition of the Juratrias, but the granitic rocks were intruded somewhat later.   Toward the end of the Neocene a period of volcanic eruptions began, consisting mainly of rhyolite and its tuffs.   After a considerable interval, during which the rhyolite lavas were much eroded, volcanoes along the summit of the range began to pour out masses of moderately basic lavas such as andesite.   When volcanic activity ceased, practically the whole of the Colfax quadrangle may have been covered by andesite to a depth of from a few hundred to over a thousand feet.   A few points, such as English Mountain, Black Mountains and Signal Peak probably remained above the surface.   Pleistocene erosion has removed the greater portion of the volcanic covering.   Glaciers, which covered a large portion of the summit region of the Sierra Nevadas during the latter part of the Pleistocene period, have left moraines in places 200 feet deep but frequently spreading out and forming only thin coverings.   Only the heavier glacial detritus which completely covers the ground has been mapped by the U. S. Geological Survey as moraines.

The Blue Cañon formation, a member of the Calaveras group, consists of clay slates and quartzitic sandstones.   The clay slates are black and fissile.   The quartzitic sandstones are dark gray and, as a rule, fine grained.

The Sailor Cañon formation is of Juratrias age, and consists of black calcareous fissile shale, interbedded with strata of quartzite and limestone.   The dip ranges from 50° to 70° ENE.   The formation is strongly metamorphosed near its contact with granodiorite.   The diabase-porphyrite and diabase are dark green and medium to fine-grained; the porphyrite usually carries crystals of dark green augite frequently altered to uralite.   The granodiorite is of light-gray color and medium-grained texture; its constituents are white feldspar, dark-green hornblende, black biotite and some gray quartz.   Near the contacts it is very common to find the ferromagnesian silicates increasing in quantity, changing the rock to diorite or quartz-diorite.   The usual form of the andesite is a tuff breccia consisting of angular or sub-angular fragments of andesite cemented by a dark-gray material chiefly consisting of finely ground-up andesite.   The andesite is a rough and porous rock of dark-gray to dark-brown color.   Porphyritic crystals of plagioclase feldspar are invariably present, as are also crystals of augite and hypersthene.   Hornblende is less abundant, but appears in many rocks as small, black, glistening needles.   Biotite is of very rare occurrence.   The ground mass in which these crystals are embedded has a structure varying from glassy to microcrystalline.

### Mineral deposits.

A series of veins is exposed between Summit City and Old Man Mountain which is very different in character from the typical gold-quartz veins of California. They strike northwest and dip 80° NE., being fissure veins, often without well-defined walls. The ores consist of pyrite, arsenopyrite, and zinc blende, carrying a moderate amount of gold. Free gold occurs only in the decomposed surface material. The gouge consists of black tourmaline and epidote. Frequent attempts made to exploit these veins generally have not been successful, owing to their low grade and absence of free gold. The wall rocks are grano-diorite, diorite and diabase-porphyrite.

A vein similar in character to the veins of Summit City is exposed at Jackson Lake, north of English Mountain, in diabase-porphyrite. It strikes northwest and dips 50° SW. The decomposed portions contain considerable free gold, and the vein was successfully exploited during 1895 and 1896.

A short distance south of Signal Peak a few small quartz veins have been prospected, apparently with little success.

A few poorly defined veins carrying auriferous sulphides have been found at the outlet of Meadow Lake. Frequent attempts made to work these have been unsuccessful on account of their isolation and low grade, together with the severe climate.

In Rattlesnake Creek nearly south of Signal Peak, copper pyrites accompanied with cobaltite occur principally as impregnations in gneissoid schist. Two or three shafts sunk at this point have not developed ore in commercial quantity.

## NEVADA CITY DISTRICT.
### Including Banner Hill District.
#### Location.

This district, surrounding the county seat, has an average elevation of about 2500 feet and is situated only four and one-half miles north-east of Grass Valley, from which place it is reached by a 20-minute ride on either steam or electric cars, the latter running every hour. The county seat has a population of about 2500, and is the principal rail and supply point for the mining camps of Alleghany, Forest, Downieville, North Bloomfield, Washington and other places in Nevada and Sierra counties.

#### Topography.

Banner Hill District lies east of Nevada City. Two long ridges, Harmony Ridge at the north and Banner Hill-Town Talk Ridge at the south, traverse the two districts in a general east-west direction. These ridges are highest at the east, Banner Hill reaching an elevation of 3904 feet. Both drop away to the west, to a rolling plateau. The town lies between them, on Deer Creek, the principal stream, a tributary of Yuba River. This creek has a rather steep grade, from 100 feet to 150 feet to the mile, in the eastern part of the area. Town Talk Ridge is the divide between the tributaries of Yuba and Bear rivers.

### History.

While some placer mining was done along Deer Creek on and near the site of Nevada City in 1849, the camp was not named until the spring of 1850, when the first alcalde was elected and the first hotel was opened. The alluvial placers were rich and the camp grew fast, having nearly 1000 houses in 1856. Both sluices and the hydraulic giant were perfected, if not first used, in this district in the diggings near Nevada City, the hydraulic process having been employed as early as 1853 by E. G. Matteson on American Hill. With the gradual exhaustion of the richer shallow placers, the beginning of work on the large but leaner deep placers of the old river channels called for the employment of great volumes of water under high pressure, and hydraulic mining flourished in the twenty years previous to 1880. Some hydraulic work was done on ground carrying the decomposed angular outcrops of quartz veins. North of the town the Manzanita Channel was worked from both ends and in the extreme northeast and northwest corners of the district and along Deer Creek near town, smaller operations were carried on. In many places the channels are so deeply buried as to be beyond the reach of profitable operation by hydraulic mining. Southwest of Sugar Loaf Mountain, which lies north of town, hydraulicking was going on extensively in the middle sixties in diggings known as American Hill, Wet Hill, Lost Hill, Coyote Hill and others.

Where the channels were too deeply covered or where the ground was rich enough to warrant, drift mining, which began in the early fifties, was prosecuted more or less continuously until recent years. Several well defined channels were exploited in this way. The Harmony Channel on Harmony Ridge in the northeastern corner of the Nevada City District and the northwestern part of the adjoining Banner Hill area, was mined profitably through the West Harmony, Harmony and Cold Spring inclines, and the Cold Spring and other tunnels. Several of its branches were also tapped. The Harmony, worked in early days, was active as late as 1900. The ground here and in the near-by West Harmony, as far as accessible, is thought to be about worked out. The gravel in this channel was cemented, requiring stamp mills, and the bedrock was swelling so that pretty close timbering was necessary. The cost of milling at the Harmony was said to be only 13 cents a ton in 1896 and total cost of mining and milling was stated to be 90 cents a ton, with the gravel carrying from $1.25 to $2.50 a ton across a width of 175 feet and a height of four feet, and with some richer streaks going as high as $4 a ton.

The Manzanita mine on the north-south channel of the same name about a mile north of town, was located in 1850 and yielded well both as a drift and hydraulic mine. Its output under the latter process was said to be $1,500,000. Drift mining in the sixties was said to have yielded $3,500,000 according to some reports, with much coarse gold. It was worked as late as 1894. The Odin and Nebraska inclines were on the same channel to the north.

Other drift mines, now idle, have been opened in a channel having a northwest-southeast direction and known as the Cement Hill Channel.

The first discovery of a gold quartz vein was made in 1850, when the Gold Tunnel vein was found. This property was worked quite steady up to 1875. It is credited with a production of $300,000 up to 1855 and about $1,000,000 in all is said to have come from the California and Gold Tunnel workings on the same vein.

The Providence and Merrifield mines on Deer Creek and the Pittsburg vein were found in 1851. The early followers of quartz mining here met the same difficulties encountered at Grass Valley. However, important production began in the early sixties. The Pittsburg is credited with $200,000 production to 1862, and a total of $1,100,000 to 1879; the Lecompton $220,000 to 1863 and the Sneath and Clay $180,-000 in 1862 and 1863. Statistics for 1865 credit the quartz mines of the district with a yield of $400,000 and for 1866 and 1867 a half million dollars yearly. The Banner and Soggs mines were also yielding good ore at this time.

The Providence mine has been and remains the best producer in the district. The early operations were hindered by the high percentage of sulphides in the ore and success awaited the use of the chlorination process about 1870. The shaft was 1100 feet deep in 1886, and besides the workings on the Merrifield vein a crosscut had been driven to the Ural vein. Eight years later apex litigation with the Champion Company began and dragged on until 1902, when the Champion Company bought the Providence. The latter is said to have produced about $5,000,000 previous to this. The vein in the Providence is large and the ore shoot has persisted from near the surface to an inclined depth of over 2700 feet, from which latter level stoping of good ore was going on in September, 1918. The Champion alone has produced about $3,000,000.

The Champion group now controls 440 acres and includes the Champion, Cadmus, Home, Merrifield, New Year, Nevada City, Providence, Wyoming, Soggs and several less celebrated claims, having a total estimated production of from $8,000,000 to $20,000,000. The property is now the only producer in the district, not necessarily because it is the best, but because it is owned by the strong North Star Mines Company, which has been able to prosecute vigorous work in recent years, while other mines of good reputation and undoubted merit lie idle for want of capital.

### Geology.

For a complete discussion of the historical and structural geology of the district, the reader is referred to the various publications listed at the end of this section. Mining activity since the publication of Lindgren's work on the gold quartz veins of Nevada City and Grass Valley districts, has not been sufficient at Nevada City to throw much new light on the geology of the region.

The following brief introductory notes are intended to give the reader an idea of the principal veins, the work done upon them and the characteristics revealed which emphasize the difference between these veins and those at Grass Valley.

#### VEINS OF BANNER HILL REGION.

The veins here are generally narrow, and most of them strike east with dip either north or south at a low angle. Exceptions are the Big Blue and Independence veins of the Murchie mine and the New York vein of the Texas system. The Big Blue vein is 4 to 5 feet wide and stands nearly vertical; the New York vein is 3 or 4 feet wide. These east striking veins owe their origin to the mineralization of two sets of fissures or sheeting planes noticeable in the granodiorite.

The Bellefontaine vein is from 4 inches to 2 feet wide, averaging about one foot, and has developed numerous small but rich ore shoots. It is in granodiorite. No recent work has been done on it.

The Federal Loan vein averages about one foot in width and has clay slate walls, somewhat altered by the intrusion of the near-by granodiorite. It carries as much as 6 per cent of sulphides, with pyrite and arsenopyrite abundant and galena, zincblende and chalcopyrite in lesser amount. The vein has been worked to an inclined depth of 1000 feet and an ore shoot 200 feet long was found. Ore is said to have averaged $15 a ton.

The Lecompton vein, credited with a production of about $250,000 up to 1867 but idle most of the time since, is from 4 to 8 inches wide in granodiorite walls. The ore is said to have averaged $40 a ton. It contained arsenic and antimony, and about equal amounts of gold and silver by weight in some parts of the mine.

The Deadwood vein had a maximum width of 18 inches. It was developed by an inclined shaft 500 feet deep, with drifts. The ore was rich, and contained a high percentage of sulphides. Total production about $300,000.

The Murchie veins have been developed by a 1150' shaft on the Murchie or Big Blue vein, which is in granodiorite and has a width of 4 or 5 feet. The walls are hard and unaltered. The ore carries considerable silver and 2 per cent sulphides. The other three veins of the group, called the Independence, Lone Star, and Alice Ball, are only slightly developed, but the property is thought to be a promising one. Ore produced between 1902 and 1910 averaged $20 a ton, and the total yield is said to have been $1,150,000.

The St. Louis vein has been superficially worked on the Alpine property at the east and on the Sharpe property at the west. It is strong and persistent, but has proven of low grade where opened. It has been traced a distance of 7000 feet, and the width, including altered formation, reaches 12'.

The Orleans-Glencoe vein has been traced $2\frac{1}{2}$ miles, half in the Banner Hill District and half in the Nevada City District. The strike is nearly west and the dip 70° to 85° S. It lies in slate walls the entire distance. The fissure zone of this vein reaches a width of 12 feet in the Sharpe property, but the quartz vein proper is from two to four feet wide. Here it has yielded ore worth from $6 to $12 a ton. On the Glencoe claim, a mile south of Nevada City, the vein has been opened to a depth of 100 feet, but shows no rich ore. The same vein has also been prospected in the Gracie shaft 100 feet deep, in the Orleans shaft 200 feet deep and in the Fortuna shaft 250 feet deep, but no valuable ore shoots have yet been found on it.

The Canada Hill vein occurs in the southeastern part of the Nevada City granodiorite area. It strikes north and dips west at a small angle, being at times nearly flat, and varies in width, not exceeding 18 inches. On the Canada Hill ground it has been opened by an inclined shaft 1500 feet deep and about 2 miles of drifting has yielded good ore. The total production is not known, but 8 years' work gave about $360,000. The silver content of the bullion was 27% and the vein showed ribbon structure of the sulphides with arsenopyrite, galena, zincblende and pyrite present to the extent of about 3%. The vein is faulted several times, the displacement being 150 feet on the surface or 45 feet vertically, where the big, low-grade St. Louis vein crosses.

Another vein system called by Lindgren the Mayflower complex, lies just south of Canada Hill in the clay slate. While this system contains eight well-defined veins, very little work at depth has been done on any of them. The Butterfly, North Star and Big Blue veins are east striking and nearly vertical in dip. They fault the flat, north striking veins, the Beckman, Floyd and Mayflower, with a number of small step-like breaks. The Beckman vein has been mined extensively on the surface. It was also opened years ago by means of an adit 800 feet long which reached a depth of 215 feet, and by a shaft 700 feet deep from which, on the 600-foot level, 1000 feet of drifting was done. In 1915 a vertical shaft was started, and some ore from old workings was milled. The vein as mined in the 90's was about one foot wide, rich in sulphides and carried some coarse gold high in silver. The Grant vein was opened here and also on the adjacent Canada Hill ground and yielded some good ore.

The Banner vein in slate was worked in the old Banner mine west of the hill of the same name, and its four northern extensions, called the Dunninton, Reindeer, Tinny and Woodville, were worked in the North Banner mine. The Banner is a stringer vein about four feet wide. It yielded over $135,000 to the middle of June, 1871, but little since. The sulphides were very rich and the ore was said to yield $20 to $30 a ton.

The North Banner veins, of which the Woodville was the chief producer, occur in diorite. The production from 1889 to 1892 was $175,000 and the last bullion was obtained in 1896. An inclined shaft reached a depth of 500 feet on the dip of the Woodville, which averaged 2 feet wide and carried high percentages of sulphides and silver.

### VEINS OF NEVADA CITY DISTRICT.

The Reward-Gold-Tunnel-Oustomah vein has been traced 7000 feet from the Reward shaft, 3000 feet southwest of Nevada City, to the Oustomah on the north, where it is covered by auriferous gravel. It lies entirely in granodiorite, dipping 30° in the California but flattening out in the Oustomah shaft from 45° to less than 20° in the deeper workings. The strike also varies through an arc of nearly 90° between the known extremities of the vein, from north-northeast to north-northwest.

This vein was the first to be mined in the district. At the Reward shaft the vein is split in two. Work done in 1894 to 1896 showed a 16" vein carrying 5% sulphides said to be rich. The shaft was 400 feet deep in 1896 and sulphide ore was being shipped, but apparently not much has been done since. In the California mine, north of the Reward, a tunnel to the vein reached a depth of 700 feet with the vein showing from one inch to four feet wide and said to carry $19 ore. The production of these two properties is unknown. The Gold Tunnel workings, begun in 1850, are north of Deer Creek. From 1852 to 1855 this mine produced $300,000 from a vein averaging 14" wide, and said to have carried ore worth $50 a ton. Work continued quite steadily until 1875. The Gold Tunnel and California, now consolidated, are credited with $1,000,000 production. The most northerly and most recent work on the vein was at the Oustomah, formerly called the Pennsylvania. Here the greatest depth on the vein was attained, and it showed those characteristics which distinguished the veins of this district from those of Grass Valley. The walls are from 3 feet to 6 feet apart, enclosing a vein of quartz one to two feet wide, with the remaining space filled by altered granodiorite. The sulphide content is high, being from 5% to 8% of pyrite, galena and zincblende. At times large cubes of pyrite, nearly an inch across, are said to have been found carrying wires of gold. This is unusual, as such coarse pyrite is usually barren. The vein here strikes N. 25° W. The shaft had an inclined depth of 1045 feet, with the dip of the vein decreasing to 20°. The vein splits again in this mine, one branch going southwest, the other north. The production is unknown.

The Mountaineer vein lies between Nevada City and the Champion group. It has been traced 5000 feet, having a northerly and northeasterly course and dipping 38° E. It has been opened in the Mountaineer mine to a vertical depth of nearly 1000 feet by a tunnel driven from the north bank of Deer Creek and an inclined winze, and was drifted on for a total length of 3700 feet. The vein pinches and swells, varying from a seam to 10 feet in width, but the wider portions are barren or low grade. The granodiorite walls are usually hard and unaltered close up to the vein, indicating little movement. The quartz in the pay shoots carried free gold with much silver and 3% or more of pyrite, galena and zincblende; ore was said to average $15 a ton. The massive quartz away from the shoots is said to carry $1.50 to $2.00 a ton. The vein splits a short distance north of the main ore shoot (which was about 1100 feet in the tunnel) and no other important ore bodies were reported from this direction. Total production between $2,000,000 and $3,000,000. Last work was done in 1916.

The Merrifield and Ural veins are now controlled for a distance of 8000 feet by the North Star Mines Company. These are the most important of the group of veins found in the system about a mile west of Nevada City, that were opened first in a number of mines named previously in this chapter, all of which are now consolidated as the Champion Group. These veins are discussed so fully in the description of the Champion Group that it is only necessary to mention a few recent developments. Since the body of this report was written activity has been largely transferred to the Providence ground. The Providence shaft has been sunk to a depth of 2700 feet, following the Merrifield vein all the way with a dip of 38° to 40°. This vein remains in granodiorite and continues to produce good ore. The ore shoot was being stoped in September, 1918, from the 2700-foot drift. This persistent ore-body here shows a length of 300 to 400 feet and is 2 to 10 feet thick. A winze started from the 2700-foot level had reached 2800 feet and drifting had started from there. Another winze sunk from the 1600-foot north drift had reached 1800 feet, from which level a drift struck good ore. This winze will be continued to meet a raise from the 2700-foot level. The Champion shaft in recent years has also reached a depth of 2700 feet (inclined), but the Ural vein has evidently not come up to expectation. It is said the Champion ore-shoot was bottomed at about 1200 feet. Ore recently mined (1915) from the Ural vein has come mostly from the Nevada City ore shoot which was connected with the Champion shaft at the 1000-foot level by a drift nearly a mile long. This shoot on the latter level showed a thickness between walls of 2 to 4 feet and in places the quartz was thickly matted with sulphides of good grade. The Champion and Nevada City shafts have in the past two years become auxiliary shafts and the Providence shaft is used for hoisting ore.

The ore from this group of mines carries no specimen gold. The sulphides average 6% or more of the ore and carry 30% of the value. About 80% of the gold is saved by cyaniding, the balance by outside amalgamation. This will probably lead ultimately to the introduction of the flotation process, which has been tested here, and is thought applicable to the ore.

The Pittsburg vein, striking N. 45° E. and dipping 43° SE.; and the Gold Flat or Potosi vein, with northerly strike and dip of 40° E., are now controlled by the Pittsburg-Gold Flat Company. The shaft has been put down to a depth of 1625 feet on the Pittsburg vein. The only work of importance since the body of this report was written has been an attempt to find the Gold Flat vein from the Pittsburg shaft. On the surface the veins are about 1000 feet apart, and it was thought to be a simple matter to connect them underground by crosscutting along the line of one of the post-mineral faults. It is reported that a 1500-foot crosscut from the 1300-foot level of the Pittsburg shaft toward the Gold Flat failed to disclose the latter vein.

**Bibliography.**

Reports of the State Mineralogist, from 1886 to 1896. Folios of Geologic Atlas of the United States, U. S. Geological Survey, Nos. 18, 29 and 66. Seventeenth Annual Report, U. S. Geological Survey, part II, pages 13 to 262. The Gold Quartz Veins of Nevada City and Grass Valley Districts, by W. Lindgren. U. S. G. S. Prof. Paper 73, Tertiary Gravels of the Sierra Nevada, by W. Lindgren, 1911. U. S. G. S. Mineral Resources of the United States for the Years 1905 to the Present.

## NORTH BLOOMFIELD MINING DISTRICT.

**Including the region about North Bloomfield, Columbia Hill, Malakoff, Relief, Lake City, Snow Tent, Moore's Flat, Orleans, and Snow Point.**

This district occurs in the central part of Nevada County, 14 miles northeast of Nevada City. The nearest shipping point is Nevada City, terminus of the Nevada County Narrow Gauge Railroad, 14 miles by a fair road which crosses the South Fork of Yuba River at a point 5 miles southwest of North Bloomfield. The elevation of North Bloomfield is 3200'.

The district occupies the divide between the Middle and South Forks of Yuba River. The climate is temperate with dry, warm summers, and heavy rains with some snow in winter. The lumber industry is important since the region is within the forest belt of the Sierra Nevada. There is an abundance of yellow and sugar pine, spruce and fir, with oak at lower elevations.

Gold is obtained extensively in Tertiary river gravels and the production is reported to have been $3,500,000 up to the year 1900.

### History of mining.

Hydraulic mining has been carried on at North Bloomfield on a very large scale. Excavations from 500' to 600' in width, extend for 5000' and reach a depth of 500'. The deposit has been opened by a bedrock tunnel 7874 feet long with entrance in Humbug Cañon. This tunnel together with other preliminary work is said to have cost $3,000,000. Soon after its completion hydraulic mining was hindered by anti-debris legislation and only such gravels have been worked whose tailings could be impounded before reaching the river.

The gravel produces on the average of 4¢ to 10¢ per cubic yd., the richest portions lying near bedrock. The yield between the years 1866 and 1900 was approximately $3,500,000, from the 30,000,000 cu. yds. excavated. It is estimated that 130,000,000 cu. yds. remain. A similar yardage occurred to the west, near Lake City.

At the Derbec mine, one mile due north of North Bloomfield, a shaft and workings have exposed a deep channel extending several thousand feet eastward. This channel connects with the main one of the region and has been mined for 7000 feet upstream from the shaft. The mine was operated from 1877 to 1893, and the production often reached $200,000 per year.

Hydraulic work was formerly carried on at Relief, where drift mining is now being pursued. The Union tunnel, 2500' long, is reported to have yielded from $30,000 to $90,000 annually for a number of years.

Water is supplied by the North Bloomfield ditch, carrying 3200 miner's inches from Bowman Lake.

### Bibliography.

Lindgren, W., Tertiary Gravels of the Sierra Nevada, U. S. Geol. Surv. Prof. Paper No. 73, pp. 139–141. Min. Res. of the U. S., 1905, p. 179; 1906, p. 191; 1907, pt. 1, p. 215. U. S. Geol. Surv. Topographic sheet of Colfax. U. S. Geol. Surv. Folio No. 66, 1900.

### Topography.

The district is on a broad, comparatively flat divide between the Middle and South Forks of Yuba River. The divide is covered with andesite and characterized by elevated level areas such as Moore's Flat and Relief Hill. The descent to the rivers both north and south is steep and abrupt in places, the average slope being from 1 in 2 to 1 in 3. The region is drained by the North and South Forks and tributaries of Yuba River.

### Geology.

Metamorphosed sedimentary rocks of the Delhi, Cape Horn, Relief, and Blue Cañon formations occur from west to east as broad. northwest-southeast bands. The Delhi is the more extensive formation, making up half the width of the southern end of the belt. Amphibolite extends in a north-south belt, about a mile wide through the center of the district, between the Delhi and Cape Horn formations. At the southern border of the district this belt resembles the fingers of a hand, the space between the pointed amphibolite fingers being occupied by Cape Horn slates; only the little finger, greatly elongated, extending south as a belt ¼-mile wide. Another lens-shaped body of amphibolite extends across the Relief cherts and quartzite and into the Blue Cañon formation. Small bodies of diorite intrude the Delhi formation to the northwest; one of these occurs as a long, narrow, curved band between the Delhi formation and amphibolite to the east. A small lenticular shaped intrusion of granite occurs in the Cape Horn slates about a mile east of the amphibolite belt. The Relief formation is composed of fine-grained quartzite alternating with siliceous gray slate. The general strike of the strata is from north to south, with dip nearly vertical; in detail, the stratification planes are exceedingly crumpled. Many small irregular veinlets and bunches of white quartz occur in the quartzite. The rocks of the Blue Cañon. Cape Horn and Delhi formations are further described in the reports on the Alleghany and American Hill districts, Sierra County.

The greater part of the region, especially the central part, is covered by andesite breccia, which overlies Tertiary river gravels in many places.

Diorite is a medium-grained granular rock composed chiefly of feldspar and hornblende, the amount of hornblende being equal or exceeding the feldspar. Hornblende in the diorite is of a black to dark green color, while feldspar is the light-colored constituent. Amphibolite is a schistose rock of fine-grained texture derived from dioritic rocks by pressure.

### Mineral deposits.

No gold quartz veins of economic importance have thus far been reported.

Gold is found in extensive Tertiary river gravels, which are continuous with the deposits at North Columbia and Badger Hill, in the adjoining Smartsville quadrangle. Gravel is found at Lake City, North Bloomfield, Moore's Flat, Orleans, and Snow Point, possibly representing an extension of the channel from Grizzly Hill (2½ miles south of North Columbia) to Badger Hill (in the Smartsville quadrangle). A channel passing through Relief is probably the one which occurs at Omega and Alpha, in the Washington district. This joints the North Bloomfield channel, east of Columbia Hill, 1¼ miles north of the town by that name.

The bedrock in the region about North Bloomfield, consisting of the Delhi formation, rises both north and south of the main channel. The channel bed is level for nearly 400' across; the deepest blue gravel is 130' thick overlain by heavy-bedded, light-colored sand and clay sometimes 100' thick; this is interstratified with fine gravel, and with andesitic tuff near the top. Unconformably above these occur 600' of tuffaceous breccias. Most of the gold occurs in blue gravel, the richest parts being close to bedrock, but owing to the great width of the channel the gold is not concentrated sufficiently to make drift mining profitable.

The Derbec channel, one mile north of North Bloomfield, is part of that which occurs at Relief. The pay gravel at the Derbec mine was from 150' to 600' wide, and 8' to 16' deep, with an average value of $2.47 per ton. The coarse gravel contains many granite and other boulders.

At Relief this same channel reoccurs as a flat terrace on the south side of Relief Hill, and on the north side of the cañon formed by the South Fork of Yuba River. The gravels fill a deep trough in a bedrock of Cape Horn slates and cover about 200 acres. Above the terrace, andesite breccia covers the region; below, the bedrock slopes down to the Yuba River. The oldest gravels, which are coarser and contain less quartz, are 60 feet deep, and covered with from 100' to 200' of alternating sands, fine quartz gravel, and clay.

At Orleans and Snow Point small areas of auriferous gravel occur and hydraulicking has been carried on. The amphibolite bedrock rises rapidly to the south. Drift mining has been carried on only at Snow

Point where the gravel bank is 135' high; the lowest fifteen feet is coarse gravel, which is in turn overlain by 20' of clay.

At Moore's Flat, 1½ miles southwest of Orleans and 4 miles northeast of North Bloomfield two bodies of gravel are exposed. An eastern one rests on a bedrock of amphibolite, while one to the west rests on Cape Horn slates. Andesite tuff covers the gravel to the south which is similar to that at Snow Point. It varies in thickness from 100' to 130'. Quartz boulders from 2' to 6' in diameter are often found on the bedrock. In 1900 it was estimated that 26,000,000 cubic yards of gravel had been washed away and that 15,000,000 cubic yards remained.

The only known quartz vein in the region is a fissure vein in amphibolite occurring one-half mile southwest of Orleans.

A mass of pyrite containing copper occurs in amphibolite on Humbug Creek, about one-half mile above North Bloomfield. Sufficient development work has not been done to determine the value of the deposit. At several places south of North Bloomfield chalcopyrite is said to occur disseminated in amphibolite.

## NORTH COLUMBIA MINING DISTRICT.

The North Columbia district is known for its extensive deposits of Tertiary river gravels, and for the productive Delhi quartz vein which occurs in its northern part. It is situated near the northwestern border of Nevada County, 8 miles in a straight line and about 14 miles by a good, winding road north of Nevada City. The nearest depot on the Nevada County Narrow Gauge Railroad is Nevada City.

North Columbia lies at an elevation of 3000 ft. The summers are dry and warm, while the winters bring a heavy precipitation of rain with snow. Yellow and sugar pine, spruce, and fir are plentiful. Water is supplied by the ditch of the Eureka Lake Company and by the neighboring streams from Fancherie and French lakes.

As in the other districts in this region gold is the only mineral product of present importance.

### History of mining.

At North Columbia the auriferous gravels are developed to a greater extent than at any other place. They are owned chiefly by the Eureka Lake Company, whose claims in 1900 covered an area of 1445 acres along 2¼ miles of channel. Much surface work has been done and 150' of gravel has been washed. The Delhi quartz vein was worked prior to 1893, after which it remained idle until 1898, chiefly because a great amount of water was encountered below the tunnel level. The mine was again opened up in the year 1898. The Gothardt vein has been developed by a vertical shaft 380 feet deep.

**Bibliography.**

Lindgren, W., Tertiary Gravels of the Sierra Nevada. Prof. Paper No. 73, U. S. Geol. Survey, page 139. U. S. Geol. Survey Min. Res. 1908, pt. 1, p. 341; Colfax Folio No. 66, 1900.

**Topography.**

The North Columbia district is continuous with the North Bloomfield district to the east, and forms part of the divide between the Middle and South Forks of Yuba River. North Columbia lies at an elevation of 3000 feet on the western side of Spring Creek. Northeast of the town the surface rises gradually to the level-topped Columbia Hill at an elevation of 4200'. The upper waters of Grizzly Creek flow parallel to the Middle Fork of Yuba River, between Columbia Hill and Grizzly Ridge. Grizzly Ridge is made up of a series of low east-west hills, about 3300' in elevation; it constitutes a divide between Grizzly Creek and the Middle Fork of Yuba River.

**Geology.**

The Delhi formation, consisting of black, or dark brown siliceous metamorphosed sedimentary rocks, is intruded by diabase, diorite and gabbro, and covered by a cap of andesite breccia northeast of North Columbia. The diabase is exposed as long, slender lenses, trending in a northwesterly direction through the central portion of the district, apparently continuous with the diabase of the Pike district to the north. Diorite occurs about ½ mile east of North Columbia, also along Spring Creek to the south and near Edwards Bridge in the southwest part of the district. Gabbro occurs along Grizzly Ridge, cutting across the Delhi formation and the diabase.

The Delhi formation is the oldest in the district and was folded and disturbed along with the other sediments of Carboniferous age at the beginning and end of the Juratrias period. The igneous intrusions occurred in early Cretaceous time and the filling of the various fissures by vein material is supposed to have followed. The gravels which are now exposed in the district accumulated during the Tertiary period of erosion. At the end of Neocene, Neocene-Pliocene time, an extensive flow of andesite breccia covered the region and solidified, covering and preserving many of the auriferous gravels. During the following Pleistocene erosion period most of this andesite was worn away; considerable remains in the region northeast of North Columbia.

**Rocks.**

The diorite of North Columbia and Edwards Bridge is a medium-grained, granular intrusive composed of about equal amounts of light-colored soda-lime feldspar and black or dark green hornblende. It is probably part of the body of diorite exposed at the head of Grizzly Creek, north of Columbia Hill.

The gabbro on Grizzly Ridge is an extension of that in the Smartsville quadrangle to the west. It is coarse-grained, being of soda-lime feldspar, diallage, hypersthene, and olivine, and is of a dark green or black color. Near the contacts finer-grained varieties occur which in part grade into diorite.

The diabase is porphyritic, being composed of soda-lime feldspar and black or green-black hornblende or pyroxene, olivine and biotite may also occur. Where exposed it is dark green in color and of altered appearance. The fresh rock is also fine-grained and the pyroxene is often converted to uralite.

### Mineral deposits.

Gold is found both as placer and lode deposits. Gravels are extensive throughout the region, but lodes occur only in one group north of Grizzly Ridge.

The placer deposits at North Columbia are the most extensive of the region. The main Tertiary river channel which branches north of North Bloomfield extends in an east and west direction through the district. The gravels are continuous with those at Badger Hill, in the adjoining Smartsville quadrangle, and form a total area covering 8 square miles. About one-half mile southeast of North Columbia a channel from the direction of Dutch Flat and Scotts Flat to the southeast joins the steeper channel from North Bloomfield; this channel was deep, with but slight grade. The North Columbia gravels are from 400' to 500' deep along the center of the channel, the deepest gravel being exposed at Grizzly Hill, one mile southwest of Kennebec House. The gravel there is coarse and made up largely of metamorphic rocks; the upper bench gravels being made up of finer quartzose material. Near the surface, especially near areas of andesite breccia, heavy beds of sand and light-colored clays cover the gravels. The bedrock is a black flinty rock, of the Delhi formation, and the deepest portions of the deposit can be reached only by running long and expensive bedrock tunnels. Injunctions against hydraulic mining stopped such development work. In 1900 it was estimated that 25,000,000 cubic yards of gravel had been hydraulicked away and that 165,000,000 cubic yards remained.

The veins of the region are of the fissure type; they occur in the Delhi formation, about 2¾ miles northeast of North Columbia, on the north slope of Grizzly Ridge. The principal group consists of three veins, of which the Delhi is the most important. The outcrops of the veins are parallel, having a strike nearly due north-south; that of the Delhi vein is about a half mile in length, dipping 75° E. The Delhi vein has a rich ore shoot, containing coarse gold, opened up by tunnels. Other veins in the vicinity are the Gothardt and the Live Oak. The Gothardt outcrop cut across the contact between the Delhi formation and diorite.

## NORTH SAN JUAN MINING DISTRICT.

### Including North San Juan, Paterson, Badger Hill, Oak Tree Ranch, Sweetland and Sebastopol.

Extensive exposures of auriferous Tertiary river gravels occurring in this district have produced a large amount of gold. An old Neocene channel extends through the district and is joined by the productive channel from North Columbia, along which large areas of gravel are exposed.

The district is located in the northwestern portion of Nevada County, with Yuba County and Sierra County bordering it on the northwest. French Corral district joins it on the south. North San Juan is 13 miles by road, in a northwesterly direction from Nevada City, the terminus of the Nevada County Narrow Gauge Railroad. Roads connect with French Corral, North Columbia and Camptonville. The greater part of the region is between 2000′ and 3000′ in elevation. Vegetation consists of yellow pine and various kinds of fir and spruce. The principal source of timber is in tracts of the Oregon Hills. North San Juan is furnished with water by the Eureka Lake Company's ditch, carrying water from the headwaters of Middle and South Yuba rivers. Shady Creek flows from Sugar Loaf Hill through the district.

### History of mining.

It has been estimated that 20,000,000 cubic yards of gravel had been excavated before the year 1891, leaving 2,500,000 cubic yards still available. Almost continuous deposits extending from North San Juan to French Corral have been worked throughout their extent, and large portions have been exhausted.

### Bibliography.

Lindgren, W., Tertiary Gravels of the Sierra Nevada, U. S. Geol. Survey, Prof. Paper No. 73, pp. 121–125, 1911. Min. Res. of the U. S., 1905, p. 179; 1906, p. 192; 1907, pt. 1, p. 215. U. S. Geol. Survey, Folio No. 18, Smartsville, 1895.

### Topography.

The topography of the district consists of low hills, between the Middle and South Forks of Yuba River, rising from the general plateau, the elevation of which varies from 2000′ to 2500′ above sea level. The area rises eastward to Sugar Loaf Hill which reaches an elevation of 3300'.

North San Juan is on the plateau in the northwestern part of the district, near the steep south side of the cañon of the Middle Yuba River. The cañon near North San Juan is 500 feet deep, while at Badger Hill (elevation of 2500') on the edge of the cañon it is 900' deep.

The southern part of the region is drained in a southwesterly direction. The waters of Shady Creek flow from the region about Sugar Loaf draining Paterson and Oak Tree Ranch, and emptying into the South Fork of Yuba River at a point southeast of French Corral.

### Geology.

The greater portion of the region is composed of granodiorite which covers an area over 4 miles wide in the North San Juan district. The belt is part of that which extends northwestward from the Nevada City region. It is a light-colored rock of uniform texture and consists of quartz, large grains of black hornblende, black mica, much plagioclase and small amounts of orthoclase. Between North San Juan and Freeman's Bridge it carries some muscovite.

The eastern contact of granodiorite with the Calaveras formation enters the southeastern corner of the district and can be followed in a northwest direction towards Paterson to a point 2¼ miles northeast of North San Juan; thence east for 1¾ miles and then northward. North and southeast of Paterson the Calaveras formation, in places, is in contact with diorite. The contact of this diorite with the granodiorite to the west is not sharply defined.

The western contact of granodiorite with amphibolite schist has a north-south direction west of Sweetland. The schistosity of the amphibolite strikes northwest and dips 68° E. At a considerable distance westward from the contact the amphibolite changes to augite-porphyrite.

The Calaveras formation is composed of closely folded clay slates and quartzites which have been crystallized along the contacts to micaceous and quartzose schists by igneous intrusions. This altered zone is rarely over half a mile in width and there is a gradual change to unaltered rock. Northeast of Badger Hill the Calaveras formation strikes northwest and dips 72° NE. It is a part of a continuous area of sedimentary rocks of the middle slope of the Sierra occurring much more extensively in the Colfax quadrangle to the east. It is the oldest formation in the district, being of Carboniferous age, and was probably folded and compressed at the end of the Juratrias period when the first great uplift of the Sierra Nevada mountains took place. The grano-

diorite and gabbro-diorite were intruded about this time, the granodiorite being somewhat younger.

The relative ages of the granodiorite, the augite-porphyrite and the
amphibolite schist have not been definitely determined, but the granodiorite of Nevada City, which is of the same period as that in the North
San Juan district is known to be later than the diabase rocks to the
south and west of it. According to the petrographic character of a
large portion of the augite-porphyrite in the Smartsville quadrangle
it should be considered as a heavy flow. The amphibolite schist was
probably formed by dynamic-metamorphism of the augite-porphyrite.
The dynamic action was intensified along two shear zones, one of which
passes through Birchville.

Auriferous gravels accumulated during Neocene erosion. At the
end of Neocene time flows of andesite breccia occurred, since which
the region has been extensively eroded. Auriferous gravels, possibly
derived from the Tertiary gravels to the east, accumulated near the
head of Shady Creek, south of Paterson, during late Pleistocene time.

### Mineral deposits.

Gold is the chief mineral deposit, occurring extensively in the
Tertiary river gravels. The Neocene Channel of Yuba River possibly
extended from Camptonville to North San Juan, and then southwest
to French Corral. About a mile north of North San Juan the North
Columbia channel, turning through Paterson, joined the main channel.
No gravels are found at the juncture, since they have been washed
away by the more recent drainage system.

The gravel deposits at Paterson and North San Juan are composed
of well-rounded pebbles of quartz, siliceous metamorphic rocks and
some sand. The gravel beds between North San Juan and French
Corral average 150′ thick, while east of Paterson they reach a thickness
of 400′. Extensive hydraulic mining has been carried on both here
and at Badger Hill, but drifting operations had not been undertaken
because the gravel was considered of too low grade. A bedrock, of
Calaveras formation, was exposed in the center of the channel at
Badger Hill in 1895. The gravel deposit west of North San Juan is
about a mile in length and rests on granodiorite. The deposit south of
this is longer and of greater width; the bedrock changes from granodiorite to amphibolite schist. None of the gravel deposits in the district
are covered with andesite breccia. The grade of the channel from
North San Juan to French Corral is 65′ per mile; the Badger Hill
channel is almost level. Three miles north-northwest of Montezuma
Hill, on the hills west of the Oak Tree Ranch, is a deposit of well-
washed gravel resting on granodiorite bedrock at a higher level than

the gravel channel of North San Juan. This deposit is probably of an earlier period than the Tertiary gravels.

In the belt of amphibolite schist extending from Birchville north to Bullards Bar several quartz veins have been found carrying auriferous iron and copper pyrites. One of the larger veins strikes north and dips 80° E.

## ROUGH AND READY MINING DISTRICT.

**Including the region about Rough and Ready, Anthony House, Rapps Ranch, and Newton.**

This district is adjacent to the famous Grass Valley district, and produced a considerable amount of gold, chiefly from gravels deposited in a branch of the Neocene channel of Yuba River. The district includes a group of workable auriferous quartz veins.

Rough and Ready is situated in the west central portion of Nevada County, 3½ miles by road west of Grass Valley, a point on the Nevada County Narrow Gauge Railroad, 12½ miles from Colfax. The town lies at an elevation of 1800 feet above sea level. The climate is mild, with moderate rainfall and some snow during winter time. The foothill region to an elevation of 2000' is generally covered by digger pine and oaks. Above 2000' yellow pine, spruce and fir predominate. The principal timber occurs as a tract in the Oregon Hills about 20 miles to the north.

There is a good water supply. Squirrel Creek, Deer Creek, and Clear Creek flow west to Yuba River. Ditches of the South Yuba Canal Company carry water from the headwaters of South Yuba River to Grass Valley for mining and other purposes. Electricity may be used for power.

### History of mining.

In the year 1891 it was reported that 3,000,000 cubic yards of gravel had been excavated at Rough and Ready and Randolph Flat, and that 1,000,000 cubic yards remained available. In 1909 the total production of the hydraulic mines at Rough and Ready, French Corral, Smartsville and Grass Valley was only $8,000.

### Bibliography.

Lindgren, W., Tertiary Gravels of the Sierra Nevada. U. S. Geol. Survey. Prof. Paper No. 73, pp. 120–124. U. S. Geol. Survey, Mineral Resources, 1907, pt. 1, p. 215. U. S. Geol. Survey Folio No. 18, 1895.

### Topography.

The region in general is broken by groups of hills, none higher than 2500' above sea level, which separate a series of parallel westward flowing streams. Squirrel Creek flows northwest through Penn Valley to a point a mile west of Anthony House, where it joins Deer Creek. Northwest of Penn Valley the ground slowly rises to hills 500' to 600' above the valley floor. Rough and Ready lies near a tributary 2 miles northeast of Squirrel Creek and 300 feet above Penn Valley. Groups of hills constitute low parallel ridges, ascending eastward to the Grass Valley plateau, about 2500' above sea level. The north side of the Rough and Ready divide is drained by Deer Creek, north of which the ground rises to a series of parallel hilltops, the highest of which is about 700 feet above the creek.

In the southwest portion of the district, south of Penn Valley, Indian Springs plateau extends eastward from Indian Springs Hill. The elevation of this plateau averages 1800'. It is drained on the north by Clear Creek, which flows northwest to join Squirrel Creek. East of the plateau the surface is made irregular by a number of low hills.

### Geology.

There is one small area of slates and quartzites of the Calaveras formation in the northeast corner of the district, being continuous with those formations of the Grass Valley region to the south. The region is composed chiefly of igneous rock. The three most extensive formations are granodiorite, gabbro-diorite, and amphibolite, occurring from west to east in the order mentioned. To the west the granodiorite is in contact with diabase; it occurs in and north of Penn Valley about Anthony House and at Rapps Ranch. It also occurs north of Rough and Ready, in Deer Creek Cañon, Kentucky ravine and on the intervening divide. Somewhat over a mile east of Rapps Ranch a body of gabbro-diorite, one mile in greatest width and four miles long, grades into the surrounding granodiorite. Southwest of Rough and Ready is a large body of gabbro-diorite, two miles in greatest width, in contact on the west with granodiorite and on the east with amphibolite. Gabbro-diorite forms the southern end of a large area of granodiorite enclosed in the main diabase area in the northwestern part of the Smartsville quadrangle. Frequent transitions from granodiorite into adjoining gabbro-diorite are found, showing that the latter is probably a magmatic segregation.

Amphibolite is in contact with gabbro-diorite both east and south of Rough and Ready. It is massive and is thought to have been derived from gabbro-diorite. The texture is not the same throughout the area, and masses of diorite and gabbro are often included. The change to amphibolite is brought about by pressure, which changes the pyroxene to green uralitic hornblende. When the change is complete the amphibolite is composed of secondary amphibol, albite, epidote, chlorite and other minerals. The amphibolite belt has a width ranging from 2 miles at the north and south ends to 1½ miles in the central portion in the neighborhood of the California and Normandie veins.

In the northeastern portion of this district, from 2 to 3 miles northeast of Rough and Ready, an area of the Calaveras formation is intruded by diabase, porphyrite, gabbro-diorite and serpentine. These formations extend northward into the French Corral district, and southward to the Grass Valley district. The diabase and augite-porphyrite resemble fine-grained diorite, dark green in color, due to partial uralitization of the augite into secondary green hornblende. The Calaveras formation is of Carboniferous age. Its sediments were folded and compressed at the end of the Juratrias period. The igneous rocks were probably intruded in late Jurassic or early Cretaceous time. Petrographic evidence seems to show that the diabase and porphyrite formed from a heavy surface flow. In the region about Grass Valley there is evidence that granodiorite was intruded into surrounding porphyrites. In the central part of the Smartsville quadrangle granodiorite and gabbro-diorite are cut by dikes of diabase and porphyrite.

During the late Tertiary period auriferous gravels accumulated, many of them being deposited contemporaneous with the period of volcanic activity. Toward the end of the period successive mud flows of andesite occurred from volcanic vents to the east and flowed down the river channels, covering the auriferous gravels. Much of the andesite has now been eroded away.

Masses of igneous rock, intruded during the Mezozoic era, were subjected to strong dynamic action, which produced schistosity in the post-Jurassic epoch of mountain building. The amphibolite of this district forms part of the zone extending from Challenge through Birchville, Newton, Wolf Creek Mountain, and Gautier Bridge.

### Mineral deposits.

Gold is the only mineral of economic importance produced in the district. It has come principally from auriferous gravels of Tertiary age. One exception is an area of recent age which occurs along Deer Creek near and above Anthony House. These may be a reconcentration of gravels from the old channel near Rough and Ready. The deposits have a bedrock of amphibolite and are partly covered by small areas of andesite breccia. The channel branches about one mile northwest of Grass Valley and a gravel deposit, partially covered by andesite, has been mined. One mile east of Rough and Ready a deposit has been mined which is uncapped and rests on granodiorite. One mile northwest of Anthony House a deposit of auriferous gravel partially exposed rests on granodiorite. Several smaller gravel areas, exposed between Mooney Flat and Rough and Ready, carry a mixture of andesite and metamorphic pebbles, probably accumulated during the volcanic period.

The California, Normandie and Seven Thirty, compose a group of auriferous quartz veins in amphibolite 2 miles southeast of Rough and Ready. The veins in general are fissures, filled with quartz by circulating waters, containing native gold, usually in rich pockets, and metallic auriferous sulphides. The veins have different dips and strikes. The Osceola vein, also in amphibolite, is one mile southeast of Rough and Ready. The Iron-clad vein fills a fissure in gabbro-diorite one and one-half miles southwest of the same town. The gabbro-diorite contains several veins rich in metallic sulphides.

In the northeastern part of the district, in the vicinity of Newton, a parallel system of quartz veins occurs in gabbro-diorite and serpentine. All of these veins have a northwesterly strike and occur in formations continuous with those of the Grass Valley district. A prospect of iron ore occurs 2 miles northeast of Rough and Ready along a contact between granodiorite and the Calaveras formation.

## SPENCEVILLE MINING DISTRICT.

The town of Spenceville is in the southwestern portion of Nevada County, 17 miles by road northeast of Wheatland, a station on the Southern Pacific Railroad.

It is situated at an elevation of about 400' and has a mild climate with warm, dry summers.

The annual rainfall of the district between October and April, ranges from about 20 inches in the valley region to 50 or 60 inches at an elevation of 3000'. The mild climate renders possible the cultivation of oranges and other semi-tropical fruits up to an elevation of 1500'. At higher altitudes the fruits of the colder climates reach their best development. The alluvial plains and valleys are used for growing cereals. A portion of the foothill region is only suitable for pasture, but some good soils are found, which, without irrigation, will yield hay and cereals as well as grapes and various kinds of fruits. Plenty of water is available from numerous small streams which flow into Bear River to the south and Yuba River to the north.

The region below an elevation of 2000' is, as a rule, covered by digger pines, various kinds of oak, and brush. Above 2000' yellow pine, fir and spruce predominate.

The mineral products of the region have been gold, silver, and copper. Most of the gold has been produced from the working of Neocene gravels near Smartsville. Some gold and silver is found in quartz-calcite veins at the Nickerson ranch. At Pine Hill an area of altered diabase contains seams of auriferous barite. The copper ores near Spenceville contain some silver, but very little gold. Deposits of cinnabar and magnetic iron occur in the district but have not been developed.

### History of mining.

The first few years succeeding 1848 were confined to gold mining in late Pleistocene alluvial deposits, some of which are now being reworked by Chinese. Hydraulicking of the Neocene gravels, formerly carried on extensively, is now confined to those mines having facilities for storing debris. Drift mining is being carried on along the bedrock of the bottom portions of these old channels. Copper ores have been mined in considerable quantity near Spenceville.

### Bibliography.

Min. Res. of the U. S., 1882, pp. 226–227; 1883, pp. 340–341; 1887, p. 76; 1907, pt: 1, p. 215. U. S. Geol. Survey Folio No. 18, Smartsville, 1895.

### Topography.

The Spenceville district, lying as it does in the foothill region, occupies a gentle slope from the nearly level lands of the Sacramento Valley to the mountains on the east. Bear River, to the south, has cut a deep channel. From Bald Rock Mountain, elevation 1674′, trending in a general northwesterly direction to the Nevada-Yuba county line, is a ridge including Lucas Hill, 1529′, Rock Mountain, 1392′, and some smaller hills with elevations ranging from 1200′ to 800′. Another ridge northeast of and approximately parallel to the last extends northwestward from Flatbort Hill, elevation 1900′, and includes Pine Hill, elevation 1800′, and Iron Mountain, elevation 1700′. Between the ridges are broad, gently sloping areas cut by stream erosion.

The principal watercourses draining this territory are the Yuba and Bear rivers. Deer Creek and its tributary Squirrel Creek, drain the northern portion of the district. The remainder of the district is drained by Bear River and its tributaries, South Wolf Creek, Wolf Creek, Little Wolf Creek, Rock Creek and Dry Creek.

### Geology.

A large irregular area of granodiorite extends from Spenceville northeastward to Indian Springs and southeastward through Sugar Loaf Mountain; on the east it is in contact with amphibolite and diabase and on the south with diabase, gabbro-diorite, amphibolite, pyroxenite and serpentine. Long narrow areas of serpentine and the Calaveras formation extend in a north-south direction through Wolf Creek Ranch, Nickerson Ranch and Gautier Bridge. About 3 miles southeast of Nickerson Ranch a mass of granodiorite is in contact with diabase. West of Mooney Flat, in the northwest corner of the district, is an exposure of auriferous gravels partially capped with andesite.

**Mineral deposits.**

Gold is found both in Neocene gravels and in quartz veins. At Smartsville extensive hydraulic mines were formerly worked, the debris being stored in an old gravel pit. Drift mining on the same channel is being carried on from Mooney Flat.

Quartz-calcite veins in diabase near the Nickerson Ranch carry gold, silver and copper. An area of diabase at Pine Hill, which alters to kaolin and siliceous rock, contains seams of auriferous barite.

Many quartz veins between Spenceville and Smartsville have northwesterly strikes and carry chalcopyrite and other copper minerals in considerable quantities. These ores are said to contain some silver but very little gold. Spenceville is the only place where such ores have been mined in quantity; one deposit being located near the contact of diabase and granodiorite is evidently a local massing of copper and iron pyrites.

One of the few occurrences of cinnabar in the Sierra Nevada is found near the Nickerson Ranch. The mineral occurs sparingly, scattered through a quartzose and dolomitic gangue along a contact between serpentine and quartzite.

A large deposit of magnetic iron ore is found four miles south of Indian Springs along a contact between granodiorite and diabase.

A deposit of limestone marked by an old lime kiln was once worked about $2\frac{1}{2}$ miles northeast of the Sweet Ranch.

## WASHINGTON MINING DISTRICT.
### Including Washington, Omega and Alpha.

The Washington district includes quartz, Tertiary river gravels and modern stream gravels. The town of Washington is located along the South Fork of Yuba River, in the central part of Nevada County, at an elevation of 2600′. It lies 19 miles northeast of Nevada City, terminus of the Nevada County Narrow Gauge Railroad. The road from Nevada City makes an easy grade along a divide south of and parallel to the South Fork of Yuba River; it has a very steep grade from the divide down to the town of Washington.

The climate is temperate, with warm, dry summers and heavy rains and some snow in winter. The lumber industry may become important, since there are large timber reserves on the ridges south of the river.

**Bibliography.**

Lindgren, W., Tertiary Gravels of the Sierra Nevada. U. S. Geol. Survey, Prof. Paper No. 73, pp. 139–141. Min. Res. of the U. S., 1905, p. 179; 1906, p. 215; 1908, pt. I, p. 341. U. S. Geol. Survey Folio No, 66, Colfax, 1900.

## Topography.

The South Fork of Yuba River flows in a westerly direction across the district; it has cut a cañon about 1300′ deep 2 miles east of Washington. The floor of the cañon is level and about ½ mile in width. Many short mountain streams from the south flow down the cañon sides

Photo No. 3. Washington, on South Fork of Yuba River. One of the older California Mining Camps. Photo by C. A. Logan.

at nearly right angles to the river, a few large streams on the north side drain the region about Graniteville; Poorman Creek, Fall Creek, and Cañon Creek. The latter drains a glacial lake to the northeast and flows along a cañon 2000′ deep, at an inclination of 40°, to the main Yuba River cañon.

### Geology.

In the central part of the district the Blue Cañon formation consists of closely folded slates and quartzites striking in a northwesterly direction and dipping steeply to the east. Washington is on the east contact of serpentine with slate; the serpentine belt is one-half mile across, near town, and widens northward. A small body of amphibolite is in contact with serpentine and slate one mile northwest of Washington; there is also a body of the same rock, in contact with slate, crossing the Yuba River Valley 2 miles west of Washington. Granite in the northeastern part of the region is in contact with slate on the west. Andesite breccia covers slates and serpentines on the divide south of the river, and an area continuous with the andesite on Relief Hill and Moore's Flat covers serpentine, amphibolite and slate in the northwest corner of the district.

Tertiary river gravels resting on Blue Cañon slates and partly covered by andesite occur at Omega, 3 miles southeast of Washington. Small areas occur to the west on slates and serpentine. Alluvial deposits of recent stream gravels occur along the floor of Yuba River Valley, both north and south of Washington. In the southeastern portion of the district glacial detritus covers areas of considerable extent.

The Blue Cañon slate is the oldest rock in the district, being of Carboniferous age. It was intruded by basic rock, and later by granite at the beginning of Cretaceous time. Tertiary river gravels accumulated during Neocene time, after which an extensive flow of andesite breccia covered the entire region. Pleistocene erosion left a capping of andesite on the high places. A period of glaciation occurred after the various cañons had been eroded nearly to their present depth. The main body of the glacier probably extended west of the Washington district, but a tongue of ice may have extended down the cañon of the South Fork of Yuba River. The present period of erosion is a continuation of the post-Glacial epoch, and has occupied a comparatively short period of time, geologically speaking.

### Mineral deposits.

Gold, chromite, and asbestos have been produced in this district. Chromite was produced particularly during 1916–1918, on account of the war-time demand. Gold occurs in lodes, in Tertiary gravels and in recent or Quaternary river gravels.

The auriferous Tertiary river gravels occur on the south side of the river cañon, along one of the main channels of the Tertiary river system, running approximately in the same direction as the present South Fork of the Yuba, but crosses it three miles west of Washington and turns north to join the main channel near Columbia Hill.

At Alpha, about one and one-half miles south of Washington, there are 75 acres of gravel preserved.  The pebbles are chiefly of quartz, quartzite and hard conglomerate; some quartz boulders as large as five feet in diameter occur on the bedrock.  The banks are 90 ft. in height, including a 20′ bed of clay on top.  About 5,000,000 cubic yards of gravel were removed in the year 1900 and about one-fourth as much still remains.

There are several hundred acres of auriferous gravel at Omega; it lies on a bench in slate and appears to extend southeasterly under the andesite cap.  The greatest thickness exposed is 175 ft.  The bed is composed of 150 feet of auriferous gravel, covered by 6 feet of clay which is overlain by 20′ of auriferous gravel.  The lowest part of the deposit contains large boulders of granite, possibly from the region about Cañon Creek; the main body is composed of material less than 6″ in diameter, of which quartz is the most prominent.  South of Omega a small gravel flat called Shellback lies at a higher elevation; beyond this, toward the southeast, the bedrock rises rapidly and gravel is found in places along the rim.

Two miles east of Omega on the east side of Diamond Creek Cañon a small body of auriferous quartz gravel 12′ thick is exposed.

Quaternary gold-bearing river gravels occur along the South Fork of Yuba River near Washington.  A short distance to the east of Washington an extensive deposit reaches to the mouth of Scotchman Creek; this gravel consists of very large sub-angular fragments and may be partly of glacial origin, although showing concentration by river waters.  Recent gravel deposits also occur below the juncture with Cañon Creek, and in Poorman Creek, 2 miles from its juncture with the Yuba.

**Lode deposits.**

There are three systems of lode deposits.  One is along contacts between serpentine and slate.  A second is a series of fissure veins in slate, such as the Washington group.  A third, found in the eastern part of the district, is a series of fissure veins in granite.  The veins as a whole outcrop from one-quarter of a mile to a mile in a general north-south direction.

A series of fissure veins crossing the South Fork of the Yuba River 3 miles east of Washington form an important group which constitutes the Washington quartz belt.  The Washington vein, one of this group, is a nearly vertical quartz body cutting across the strike of the slates; it was worked 25 years ago and produced considerable gold.

The Yuba mine, in the cañon of the South Fork of Yuba River, is one of numerous veins occurring in granite.  A narrow streak of slate and limestone, imbedded in the granite, follows this vein for considerable distance. The vein contains free-milling coarse gold associated with galena, pyrite and pyrrhotite; it varies from 2′ to 16′ in width. The Eagle Bird vein, half a mile up the river, occurs similar to the Yuba, the ore in both cases being of low grade. Veins of this group extending northward into the Graniteville district have not been very productive.

Chromite has been mined at several places in the serpentine belt which crosses South Yuba River just below Washington. All the orebodies were very small except that opened at the Red Ledge mine, near the Nevada City road, about three miles from Washington. This property has yielded several hundred tons of a fine grade of ore.

Asbestos is being mined and milled about three miles from Washington, near the old Fairview mine, where chrysotile occurs in considerable quantity in serpentine just north of the river.

## YOU BET MINING DISTRICT.

### Including Little York, Red Dog, Hunts Hill, Quaker Hill, Scotts Flat and Galbraith.

This large productive district contains extensive Tertiary auriferous gravels. One small vein, the Orono, occurs in the extreme southwestern part of the district, but is not producing now.

The district is located in the southeastern part of Nevada County. You Bet is 7½ miles in a straight line southeast of Nevada City; the nearest shipping point is Dutch Flat, in Placer County, which lies five miles east by a winding road in fair condition. You Bet lies at an elevation of 3000′. Quaker Hill, 3½ miles to the north, has an elevation of 3500′. The region has a mild climate, the summer months being warm and dry, while the winters are accompanied by heavy rainfall and snow. Timber consists of yellow pine, sugar pine, fir, spruce and a few oak. Water is supplied by the South Yuba Company, which utilizes the headwaters of the South Fork of Yuba River and lakes in the adjoining Truckee quadrangle.

### History of mining.

Hydraulic mining formerly carried on near Scotts Flat resulted in the removal of 12 million cubic yards of gravel. At Quaker Hill, the gravel removed is estimated to have been 35 million cubic yards, 140 million cubic yards remaining in 1900. Drift mining was formerly carried on south of Galbraith, and at Red Dog, and at Hawkins Cañon, near You Bet; the deepest gravels have been hydraulicked. Considerable drift mining has been done on the claims of Niece and West, one and one-half miles to the northeast. At You Bet, on the Steep Hollow side, 47 million cubic yards of gravel are estimated to have been removed from the channel, leaving 100 million cubic yards available. Much of this can not be readily washed, because there is insufficient slope to carry off the tailings water. The diggings at You Bet have produced a total of $3,000,000, the production in 1909 being $358,600.

The gravel at Little York has practically all been hydraulicked, the production having been estimated at $1,000,000 or more.

### Bibliography.

Lindgren, W., Tertiary Gravels of the Sierra Nevada. Prof. Paper No. 73, U. S. Geol. Survey, p. 144. Min. Res. of the U. S. 1907, pt. 1, p. 215; 1909, pt. 1, p. 278. U. S. Geol. Survey Folio No. 66, Colfax, 1900.

### Topography.

The region in general is hilly and with many mountain streams. You Bet is on a comparatively level area, sloping gradually westward to Greenhorn River. The cañon of Steep Hollow Creek is two miles south of You Bet, at an elevation about 400′ lower; Little York is on a long ridge separating this creek from Bear River, which it parallels. Chalk Bluff Ridge between Steep Hollow Creek and Greenhorn River at an elevation of about 4000′ forms the divide. Quaker Hill is in the cañon of Greenhorn River which flows south and joins Bear River.

### Geology.

Metamorphic rocks of Carboniferous age lie in north-south bands. The Delhi, Cape Horn, Relief, and Blue Cañon formations occur from west to east respectively.

The Delhi formation covers extensive areas in the northwestern part of the district, where it is continuous with the formations near Harmony Ridge, and Edwards Bridge to the west. It strikes northwesterly and dips 80° E.

The Cape Horn formation is the most extensive, forming a band about four miles wide. It is continuous with the Cape Horn formation in the North Bloomfield and Alleghany districts to the North.

A long finger of amphibolite, about one-quarter mile wide, occurs between the Delhi and Cape Horn formations to the north and part of the same belt extends through North Bloomfield. Amphibolite also occurs, intruded as a narrow lens-like band, southeast of You Bet.

Intrusions of diabase occurred in the western and southwestern parts of the district during post-Jurassic time, being part of more extensive bodies south of Nevada City. Small bodies of serpentine and gabbro are associated with diabase in the southwestern portion of the district.

During the Tertiary period the country was deeply eroded and a definite river system formed not unlike the present one. Auriferous gravels accumulated and rhyolite flows occurred. The massive flows were viscous and did not extend far from their sources, but the tuffs were carried down by the streams as mud flows.

The rhyolite flows were later eroded and very extensive flows of andesite, remains of which are now found on the various ridges and divides, covered the entire region. The auriferous Tertiary gravels are more extensive than the recent ones. Neocene channels from Gold Run and Lowell Hill join in the region of Dutch Flat and the resultant channel extends northwestward through Little York, You Bet, Red Dog, Hunts Hill, Quaker Hill and Scotts Flat, joining the North Bloomfield channel near North Columbia. Gravels are exposed along this course, wherever the andesite or rhyolite capping has been eroded away.

Large accumulations of gravel are exposed at Scotts Flat and at Quaker Hill where Deer Creek and Greenhorn River have cut through the rhyolite tuff and andesite breccia covering. The gravel fills a trough surrounded by deep bench gravels. The deepest channel has been exposed by mining operations at Hunts Hill, 1¾ miles southwest of Quaker Hill, and is about on a level with the Greenhorn River tailings. At Quaker Hill a shaft sunk through the gravel showed bedrock at an elevation of 2650′ while at Scotts Flat bedrock has been found at 2770′, although this is probably not the lowest bedrock.

At Quaker Hill the pay gravel varies from 4′ to 16′ in depth; the width of the channel, containing sufficient gold to make drifting profitable is 130′. In the deep trough the gravel is coarse and cemented and is often very rich. The bench gravels are chiefly composed of fine quartz gravel mixed with sand carrying about 6 cents per cubic yard in fine gold. The gravel banks at Quaker Hill reach a thickness of 250′; there and at Scotts Flat much gravel remains, but it is difficult to get dumping ground and sufficient grade for sluices.

The bedrock of the Quaker Hill gravels is chiefly made up of Cape Horn slates. A capping of rhyolite occurs to the northeast and covers deep gravels to the northeast for 3 miles from Quaker Hill. The gravel is probably a remnant of a tributary of the main Neocene channel which passed one mile east of Galbraith, where some drift mining has been carried on. Some gravel underlying 100′ of clay has been drifted one mile south of Galbraith. Three miles northeast of Quaker Hill high bedrock is exposed, east of which are the small Red Diamond channel on the north side of the ridge, and other channels on the south side. At Buckeye Hill, one mile southeast of Quaker Hill, a small mass of bench gravel occurred that has been almost entirely removed by hydraulicking.

At Red Dog to the north, and at Hawkins Cañon to the south of You Bet a deep channel is exposed, which is apparently continuous between these two points. The channel has a slight grade, the average elevation being 2620'.

The gravel at Little York in part of the old deep narrow channel, has been nearly all removed by hydraulic mining. The channel carries a hard cement gravel, 30' to 40' thick, which is capped by a bed of fine gravel in places 350' thick and interstratified with clay and sand. Large boulders of quartz and quartzite occur on the beds of both the deep channel and the benches.

Virgin Pleistocene or Quaternary gravels buried by tailings are reported to occur in Greenhorn River near Scotts Flat southeast of Quaker Hill and at Red Dog.

A quartz vein occurs in the southwest corner of the You Bet district ½ mile north of the Nevada County Narrow Gauge Railroad bridge; the vein is along a contact between diabase and serpentine. A prospect for desseminated chalcopyrite has been opened near a serpentine-slate contact near the road from Colfax to You Bet; sulphide was found.

# MINES AND MINERALS.

## ANTIMONY.

**John Johnson** of Grass Valley has an antimony prospect near the center of the SW. ¼ of Sec. 25, T. 15 N., R. 8 E.; elevation 740'. A vein of quartz carrying pyrite and stibnite is frozen to the quartz on the footwall. It is from 6" to 2' wide, strikes N. 40° E. and pitches 80° to the southwest.

**Mohawk Claim.** Owners, Wm. McLean, Graniteville; Chas. Schmidt, Sacramento.

Location: Graniteville Mining District, Sec. ?, T. 18 N., R. 10 E., situated in Moore's Flat, 21 miles northeast by road from Nevada City.
Bibliography: U. S. Geol. Survey Folio 66, Colfax. U. S. Geol. Survey. W. Lindgren, Prof. Paper 73, page 141.

Antimony ore was thought to occur in sufficient quantity and quality to justify reopening of this mine in the latter part of 1915. An old tunnel driven 600 feet on the contact of serpentine and slate, was cleaned out, but no antimony was found during this work, and the property was abandoned after about $6,000 had been spent.

## ASBESTOS.

**Sierra Asbestos Mine.** In litigation in part, 1918. Leased to Sierra Asbestos Company, 333 Monadnock Bldg., San Francisco.

Location: 2½ miles northwest of Washington, Sec. 2 or 11, T. 17 N., R. 10 E.

The Sierra Asbestos Company is mining and milling chrysotile asbestos on claims adjoining the old Fairview mine, about ½ mile north of South Yuba River, and 2¼ miles from Washington by road. Six claims located in April and June, 1917, by F. T. Smith, L. M. Ludovice, J. H. Stark, Chas. Hellmers, and J. L. Foisie, have been leased, as have also the adverse mineral rights claimed by Eleanor Hoeft on two claims said to have been located adjoining the Fairview property on the south.

The claims are in the large area of serpentine, which trends north across the river and has a width of about ½ mile where the stream crosses. In mining the asbestos no attempt is made to sort out good material. A tunnel has been driven N. 70° E. 300 feet from a point about 750 feet east of the mill, into the serpentine. From the face of the tunnel a raise was put through to the surface 150 feet. The serpentine is mined by glory hole, and is dropped through the raise to the tunnel where it is drawn into cars. The tunnel has been driven on a 6% grade, and a go-devil with cars of 30 cu. ft. is used to deliver the rock to the mine bin. Here it is taken by a gravity tram in ¾-ton buckets to the top of the mill building 560 ft. distant.

The company has bought the Fairview stamp-mill consisting of 20 stamps of 1000 pounds each, housed in a wooden building. Power is furnished by an 88 horsepower Doak gas engine burning distillate. In

Photo No. 4. Tunnel on Sierra Asbestos Company's property near Washington. The serpentine on the slope in the background is being worked by glory hole and is dropped through a raise to the tunnel level. Photo by C. A. Logan.

season a small water right under 400 ft. head also furnishes some power. Dry crushing and screening are used throughout.

From the tram the ore goes on a grizzly, oversize being broken in a Blake-type crusher. Automatic feeders send it to the stamps, which drop 6″ with clear dies, and discharge wide open at the level of the die onto a rubber belt conveyor. Each battery is provided with an air blower which helps to get the asbestos out of the battery as soon as it is broken clear of the heavy gangue. The belt conveyor delivers the ore to a double deck shaking screen, the coarsest openings of which are $\frac{1}{4}''$. The heavy oversize on this screen is about $\frac{1}{2}''$ and consists of serpentine carrying more or less asbestos not yet fiberized, or freed from gangue. This is returned by a bucket elevator to the stamps. Fine gangue through the screen goes to waste. Both long and short asbestos which has been partly fiberized is caught up by air suction at the foot of the screen, and carried into a fiberizer. This is a closed

Photo No. 5. Sierra Asbestos Company's Mill. Photo by courtesy of John D. Hoff.

cylinder about six feet long, and about 15'' in diameter, carrying revolving blades which complete the fiberizing process. The air suction, operated by a fan, draws the fiber out of the fiberizer over an air trap where heavy gangue drops out. Thence the asbestos is drawn into a funnel shaped cyclone separator, with a central stack. Here the effect of the fan gives the light fiber a whirling, centrifugal motion. It strikes the circular walls of the separator and falls, while light dust is blown on through the stack into the open air. The draft through the stack is regulated by dampers.

The asbestos drops onto a corrugated shaking screen with 1/16 and 1/32 inch holes, which screens out any remaining sand and delivers the asbestos to a revolving trommel which grades the product according to screen size. The trommel carried two revolving shafts which carry blades to distribute the asbestos. The first material through the trommel screen is a dust product of 16 mesh to 20 mesh in size. This is grade No. 4. No. 3 is $\frac{1}{8}$ inch and weighs about 80 lb. a sack. No. 2 is $\frac{1}{4}$ inch, and No. 1 is the product over $\frac{1}{4}$ inch, which discharges from the trommel and weighs about 50 lb. a sack. From the trommel the graded product drops through hoppers into gunny sacks.

The plant is still in the experimental stage, and changes are going on continually. The present cleaning and grading equipment is thought to be capable of handling the output of 15 stamps, which crush about five tons each in 24 hours. When visited, none of the milled product had been hauled, but about 200 tons of ore had been milled and some graded asbestos was ready to ship.

A good grade of chrysotile asbestos is said to occur for a length of about a half mile north and south from the Fairview mine to the South Yuba River. The best quality and greater quantity of asbestos apparently occurs in the hard, black massive serpentine, which outcrops just north of the area being quarried. The rock being sent to the mill from the glory hole is the greasy, schistose serpentine. An inspection of this gives the impression that it contains little asbestos, but the milling process reveals considerable short fiber. In the massive serpentine, asbestos may be seen with fiber $1\frac{1}{2}''$ long. During 1917 a small tonnage of No. 1, No. 2, and No. 3, and a few hundred pounds of spinning fiber was produced by hand mining and sorting. The greater portion of asbestos noted would probably run not over $\frac{3}{8}''$ long. South of the river, there has not been any development of asbestos yet, but there are some producing chromite mines, and some small bodies of chromite have been mined in the serpentine north of the river.

The milling of asbestos with stamps does not conform with the standard Canadian practice, but the operators have taken advantage of equipment at hand and are evolving a process which will give a product suitable for use in composition flooring (of which they make a specialty), and other purposes, such as steam-pipe covering, where short fiber can be used. A drier will probably be found necessary if steady work is kept up during the winter. An electric lighting plant is to be built and a fan will be installed to dispose of dust. Hauling by truck to the railroad at Nevada City, about 20 miles, will cost $10 a ton on contract, but the company hopes to be able to do their own hauling for much less. The road would normally be passable for trucks only in the dry season.

## BARITE.

**Democrat Prospect.** Owner, William Maguire, Nevada City.

Location: Liberty Hill Mining District. Sec. 24, T. 16 N., R. 10 E. 4 miles by trail or 7 miles by road from Dutch Flat. Elevation 3450 feet.

This property consists of one lode location on the Doolittle ditch one mile north of Liberty Hill mine. There is an old road, overgrown with brush, connecting the deposit with the Steep Hollow road.

Massive barite of various colors outcrops in a dome shaped knob on the nose of one of the ridges which rise abruptly from Bear River. On the lower side of the deposit where cut by the ditch, the barite is about 15 feet wide and it outcrops for a distance of about 250 feet, rising 60 .feet above the ditch level at the highest point of outcrop. It is colored pink, grey and black but no pure white barite occurs on the surface. The fairly pure barite grades into an impure barium— aluminum silicate on the north, and at the ditch level the wall rocks are of a fine grained rock with the appearance of an altered rhyolite, carrying large crystals of barite. It is said that assays show 85% to 98% $BaSO_4$. The color has so far injured its marketability.

## CHROMITE.

**Alta Hill Chrome Mine.** A. A. Codd of Reno, Nevada, recently bought from Mau, Williams and Heidrich three acres of patented land in Sec. 22, T. 16 N., R. 8 E., about a mile and a half from Grass Valley.

Seven men were working in August, 1918, and were producing about a ton a day. Two shafts 20 feet deep had been sunk and showed lenses four and six feet wide, and open cuts exposed two other lenses of good size. Several carloads of ore were produced by Williams and his associates, ranging from 31% to 36.12% $Cr_2 O_3$. The property promises a total yield of several hundred tons. The Holseman mine adjoins.

**Baker Prospect.** D. Whildin, lessee, Grass Valley. Two men were prospecting in August, 1918, on a small showing of low-grade disseminated ore, on a 17-acre location in Sec. 26, T. 16 N., R. 8 E., a mile and a half from Grass Valley, and ¼ mile off the road to Nevada City. No shipping ore had been found.

**Codd Prospect.** A. A. Codd of Reno, Nevada, has 56 acres of patented land and 40 acres of locations in Sec. 21, T. 16 N., R. 8 E., a mile and a half from Grass Valley, on which small bunches of disseminated grey chromite ore of milling grade occur. About 80 tons of ore which had been mined in July, 1918, were too low grade to ship. More prospecting was planned, but the property was idle in August, 1918.

**Davey Property.** Owners, John Davey and Son, Grass Valley; lessees, C. W. and W. J. Jenkins, Grass Valley. The lessees had been prospecting on 40 acres of patented land in Sec. 26, T. 16 N., R. 8 E.,

¼ mile from Grass Valley, on the Banner Hill road. A lens containing about 3 tons had been dug out and prospecting was to continue, with the possibility of small production.

**Dickerson Property.** Owner, L. Dickerson Estate; lessees, L. V. Dorsey and D. R. Ridge, Grass Valley. Small bunches and lenses less than a foot wide have yielded some ore, and total production for 1918 was expected to reach 25 tons. Property contains 100 acres, patented, in Sec. 4, T. 15 N., R. 8 E., about one mile from Grass Valley on Lincoln Road.

Photo No. 6. Horse whim used for hoisting chromite ore at Hoeft Chrome Lease near Nevada City. Photo by C. A. Logan.

**Dorsey and Ridge Property.** L. V. Dorsey and D. R. Ridge have 80 acres of locations in Sec. 6, T. 15 N., R. 8 E., in the Rough and Ready district, six miles from Grass Valley, and a mile and a half off the road to Deadman's Flat. A number of shallow holes had been dug on small bunches of ore and a shaft had been sunk 12 feet on a lens 14″ wide, said to carry 40% to 52% chromite. About 50 tons had been mined but not shipped in August, 1918, and the property was temporarily idle, but the owners intended to resume work.

**Geach Property.** Owner, Thos. R. Geach, Grass Valley; lessees, C. W. and W. J. Jenkins, Grass Valley. Consists of 10-acre patented area in Sec. 25, T. 16 N., R. 8 E.; ¼ mile from Grass Valley and ¼ mile from Banner Hill road. A drift driven from a shaft 20 ft. deep showed a face of ore two feet wide, and there were 10 inches of ore for a length of 8 feet in the bottom of an underhand stope. Forty tons had been mined, but not sold, and $\frac{3}{4}$ ton a day was being produced late in August, 1918. The ore outcrops on the surface 50 feet ahead of the present work and the property may be capable of producing four to five carloads.

**Golden Gate Mining Company.** Small bunches of chromite were being mined by Baskin on 50 acres included in the patented claims of this company, in Secs. 25 and 26, T. 16 N., R. 8 E., a mile from Grass Valley on Banner Hill road. The production may total a small carload.

**Hoeft Lease.** Owner, Isabelle C. Sherman; lessee, Eleanor Hoeft, Nevada City. The property is in the SW. $\frac{1}{4}$ of SE. $\frac{1}{4}$, Sec. 11, T. 16 N., R. 8 E., about 2 miles from Nevada City by road, and extends from near Deer Creek, just below the Champion mine to the shoulder of the ridge north of the creek. Some shipping ore was sluiced off the hillside and a lens of low grade was mined on the lower slope. In September, 1918, all work was being done near the top of the ridge. An incline shaft had been sunk 47 feet and a drift and small stope had then been run 30 ft. southwest. In the face of the drift there was 10 feet of ore, but as work progressed it was said the width increased. There was a soapy gouge on the footwall side and a horse of hard fine-grained black rock in the serpentine cut off the ore at an angle to the floor of the drift and ran up toward the hanging wall. The drift has been largely in ore, and the roof of the drift was in ore for several feet near the face.

The serpentine belt carrying this and adjacent deposits is well marked, crossing Deer Creek with a northerly trend and forming prominent ridges in which low grade chromite bodies occur, on both sides of the stream. No production has been made yet from south of Deer Creek, but good sized bodies are reported. Adjacent to the Hoeft lease two other operators have opened pits and were mining low grade disseminated ore. Float in small pieces of high grade is common, but no important amount of shipping ore is reported. The gangue of the chromite in the Hoeft drift is hard, unaltered and rather vitreous as if cooled rapidly from fusion or possibly opalized and forms a blackish groundmass for the disseminated chromite, strikingly different from the usual soft serpentine. The ore is characterized by a high iron content which limits the percentage of chromite that can be obtained in the concentrate. Miss Hoeft states the concentrate so far has averaged about 32% $Cr_2 O_3$, and the iron oxide content is variously stated to be from 10% to 20% in the ore, in some cases being nearly equal to the chromite content in the crude ore. The ore averages about 12% $Cr_2 O_3$.

There appears to be a considerable reserve of ore in the ground covered by this lease but there is only a scanty outcrop, and underground work has not been carried far enough to block out an ore reserve, mining having gone forward as far as drifting.  When visited, there were only about 25 feet intervening between the Hoeft drift and the pit opened by the Nevada County Chrome Company, but while the latter operators have worked their ore body nearly to the property line, there remains on the Hoeft lease a large unexplored and promising area.

Ore from the Hoeft lease has been milled in the Champion mine's stamp mill, about a half mile distant.  In September, 1918, two men were mining and timbering the shaft.  Ore is hoisted by a horse whim in ½ ton skip.  The owner gets a royalty of $4.50 per ton of concentrate and the charge for milling, concentrating and sacking is $2.50 per gross ton.  Other costs, mining and hauling bring the total cost per ton to such a high figure that it was declared there was no profit obtainable. Production thus far had been over 300 tons of concentrate.  The percentage of recovery on these ores is low, ranging from 65% to 75%. For details of concentrating practice see Chromite Concentrating Plants.

**Holseman Mine.**  Owner, H. Holseman, Grass Valley; lessees, F. M. Pfeiffer, Grass Valley, and T. Hogan, D. Muer, Thos. Gill, and Geo. J. Hothersall of Nevada City.  The property contains 34 acres of patented land in Sec. 22, T. 16 N., R. 8 E., a mile and a half from Grass Valley. Early in August, 1918, ore was being mined at the rate of a ton and a half daily.  Ore sold during the previous month ran only slightly better than 30%, $Cr_2 O_3$.  There were some old open cuts on the property, and a shaft 20 feet deep, from the bottom of which a drift had then been driven 20 feet.  Hand drills were used in mining and a windlass for hoisting.  The property gave promise of sustained production for another season.

**A. E. Hooper** of Grass Valley was prospecting for chromite on 75 acres of patented land owned in Sec. 19, T. 16 N., R. 9 E., on the Banner Hill Road 2½ miles from Grass Valley.  The small stringer uncovered at time of visit (August, 1918) had not yet yielded any ore.

The **Mount Hill Chrome Mine** is in Sec. 13, T. 17 N., R. 10 E., M. D. M., south of Washington, near the road to Nevada City, at an elevation of 4160'.  It is owned by George Scott of Washington.

About thirty tons of 45% chrome ore had been taken in June, 1917, from a lens striking N. 10° E. and pitching 65° E.  Development work consisted of a pit 6' wide, 10' deep and 14' long, showing ore on the south face 4' wide and 5' high, opened for 5' along the ore body.  From a small area 14' farther south, along the strike of the ore body, a 6' pit about 8' square has yielded 6 tons of float chrome; clay was being worked in the bottom with hopes of striking more ore below.  No ore had been shipped.

**Bob Moscatelli** of Washington has been mining chromite on Poorman's Creek, in Sec. 1, T. 17 N., R. 10 E., 2 miles northwest of Washington. The Union Chrome Company took a lease on the property for 4 years from the spring of 1917, but have recently relinquished the lease (September, 1918). The property produced about 100 tons in 1917,

Photo No. 7. Pit opened by Nevada County Chrome Company on low grade orebody near Champion Mine. Photo by C. A. Logan.

and there are said to be about 30 tons more ready to haul. Moscatelli plans to continue prospecting. The lenses of ore are small and work so far has been superficial. The ore is of good grade.

**Pete Moscatelli** of Washington has mined some small lots of chromite in the vicinity of Washington, and has sold part of a carload to Hogan and Hothersall.

**Nevada County Chrome Company** (Morgan and Leitcher) of Nevada City is mining chromite a mile and a half from Nevada City in the E. ½ of SW. ¼ of Sec. 11, T. 16 N., R. 8 E., on the ridge just north of Deer Creek, and near the Champion mine. A pit 100 feet long and 20 feet wide had been sunk to a depth of about 20 feet in September, 1918, and work on the south end had been carried to within 25 feet of the line of the adjacent Hoeft lease. The ore is disseminated chromite in a very hard gangue. The ore body strikes north and pitches nearly vertical. It was pinching off on the south.

It averages about 15% $Cr_2O_3$ content and about 10% iron, as milled. The concentrate probably averages not over 36% $Cr_2O_3$ and 15% to 20% iron. The ore is hoisted out of the cut by skip operated by a demounted automobile engine, and is hauled nearly 3 miles to the Oustomah Stamp Mill for crushing and concentration. The pit has produced several thousand tons of ore, but will be difficult to work on account of caving in wet weather and rain will probably also curtail hauling. The bottom of the pit did not show much remaining ore and the work was near the property line on the south, so the future output will depend on new lateral development. The milling process is described under Chromite Concentrating Plants.

**Red Ledge Chrome Mine.** Owners, Williamson Brothers and Cole, Washington. The property is in Sec. 13, T. 17 N., R. 10 E., about three miles from Washington, 1000 feet higher in elevation, and 15 miles from Nevada City. High grade chromite is being mined on both the north and east slopes of a long serpentine ridge, which is the southern extension of the area containing the Sierra Asbestos property across the river. The mine is worked now through two drifts which have been driven into the hill directly below the road and which will intersect ultimately. On the north slope near the contact, which forms an angle here, an open cut yielded about 400 tons of high grade chromite in 1917. Just below the cut, a drift has been driven nearly south 150 feet in serpentine and good ore has been produced from stopes which are still being mined. On the east slope of ridge, 50 feet vertically below the road, a drift has been driven in a general westerly direction for 100 feet, and stopes from this level are being worked. It appears that the workings are all on the same irregular ore-body. The property is equipped with a gasoline engine and air compressor, which furnishes air for a hoist of 1500 lb. capacity, and one air drill. Ore is hoisted from both tunnel levels to loading platforms at the roadside. Seven men were employed in September, 1918. Ore is hauled in trucks to Nevada City at a cost of $7 a ton in dry weather, but the cost is higher in winter, and hauling over wet roads has to be done with teams.

It is hoped to make a total production of 500 tons in 1918. Five carloads have already been shipped and several more were mined at time of visit.

Chromic oxide content runs from 42% to 45%. The ore is massive and no low-grade, disseminated body has been noticed. It carries

Photo. No. 8. Red Ledge Chrome Mine showing tunnel, ore pile, and incline for hoisting ore to road. Photo by C. A. Logan.

considerable of the wine colored chromium chlorite which is sometimes present in quantity sufficient to give a load of the ore a distinctly pink-grey appearance. The chrome garnet, uvarovite, is present in smaller quantity.

The operators of the Red Ledge are developing two smaller properties, one on the same side of the river a mile distant, and the other $2\frac{1}{2}$ miles from Washington, north of the South Yuba.

Photo No. 9.  High grade chromite ready for shipment at Red
Ledge Mine, near Washington.  Photo by C. A. Logan.

**Rolph Ranch.** Owner, E. J. Rolph, Chicago Park; lessee, W. S.
Moulton, Chicago Park.  Comprises 160 acres in Sec. 28, T. 15 N.,
R. 9 E., a mile and a half from Chicago Park.  Pits sunk along the
outcrop showed four feet of milling ore with occasional small lenses of
higher grade.  The greatest depth attained was only 7 feet in August,
1918, so no estimate of tonnage was possible.

**Sweet Ranch.** Owner, John Sweet, Wolf post office; lessees, Ed Morgan, Nevada City, and Henry Yue, Grass Valley. The property is a patented one-quarter of Sec. 4, T. 14 N., R. 8 E., eight miles from Grass Valley, and a mile and a half off the Auburn road. A cut 50 feet long, 10 to 15 feet wide and 40 feet deep had produced several hundred tons of shipping ore up to August, 1918. A drift had also been run 20 feet from the bottom of the open cut. It was reported that a body of milling ore 10 to 15 feet wide had been uncovered, but it is probably too far from the nearest custom mill, at the Champion mine, near Nevada City. The haul to Grass Valley in trucks costs $2 to $2.25 a ton. Only the minimum amount of work to hold the lease was being done.

**Schmidt Property.** Owner, G. A. Schmidt, San Francisco; lessee, W. S. Moulton, Chicago Park. Some small lenses of chromite have been found on $6\frac{1}{4}$ acres of patented land in Sec. 21, T. 15 N., R. 9 E., a mile from Chicago Park and on a branch road a mile from the main road to Grass Valley. The property was idle in August, 1918, and the lessee was looking for financial help to do further prospecting.

**Thompson Ranch.** Owner, Herman Thompson, Grass Valley; lessee, Henry Yue, Grass Valley. The lease is of 640 acres, patented, in Sec. 5, T. 14 N., R. 8 E., 13 miles from Auburn, or Grass Valley, and 3 miles from the highway. A drift driven from bottom of a 20 ft. shaft shows ore $3\frac{1}{2}$ to 4 ft. wide for 17 feet on the floor and in the face of the drift. No ore had been sold in August, 1918, but about 30 tons had been mined and the faces of ore exposed as well as float indicated a body probably capable of producing some hundreds of tons. Two men were employed. Ore will be hauled to Auburn.

The **Turtledove** chrome property is in Sec. 1, T. 17 N., R. 10 E., M. D. M., at an elevation of 2780', about one mile northwest of Washington. It consists of one claim owned by Walter Niles, Fred Miller and H. O. Kohler of Washington.

Development work had been done on a spur ridge on the east side of Poorman's Creek. On the south side of the ridge about 20 tons of good ore was taken early in 1917 from a lens striking N. 10 degrees E.; it had been opened up by a cut 4' deep, 2' to 4' wide, and 25' long. On the north side of the ridge, 75' north of the other workings, a prospect hole 6' long, 4' wide and 5' deep had found only traces of chromite along a north-south seam in the serpentine. An assay of a sample of ore taken near the surface is said to have run 52.16% $Cr_2O_3$. Idle in September, 1918.

The **Woil Property** is reported to lie two miles northeast of Grass Valley. Ten tons of ore, said to average 56.68% $Cr_2 O_3$, were produced in 1917, but no more ore was found. The ore was purchased by T. F. Hogan of Grass Valley.

The **Wolf,** or **Limekiln,** chrome deposit is in Sec. 4, T. 14 N., R. 8 E., at an elevation of 1480', 14 miles northwest of Auburn. The property is owned by H. Thompson of Wolf and leased by Guy Walsh and Mr. Hall of Auburn.

A series of chrome lenses strike N. 40° W. and pitch 80° NE. The main working consists of a pit 4' wide by 8' to 10' deep. The ore exposed on the footwall is granular and carries considerable silica. The solid ore body carries a fine grained mixture of chrome and magnetite. About 30' southeast of this is another pit 3' wide by 6' long by 10' deep, in which a cross-stringer has been opened up along the southeast end; this stringer was exposed 14" wide by 4' long by 4' deep and struck in a northwesterly direction; it had been traced for 75' farther northward by an open cut from 2' to 3' deep. About one-half of the ore exposed had been mined. It is reported that H. C. Schroeder of Nevada City shipped 55 tons of 35% ore from this property in 1916.

### Chromite Concentrating Plants.

**Champion Mill.** The process of milling used here was simple and the operators have attempted no innovations or additions to the plant as they consider that present conditions and future promise of chrome mining do not warrant added expense for treatment.

During the summer and fall of 1918, two batteries of five stamps each were used part of the time to crush chromite, it being stated that the average mill capacity devoted to this use was equivalent to five stamps, three-quarters of the time.

The stamps weigh 1250 lbs. each and crushed through 40-mesh screen, dropping $6\frac{1}{2}$ to 7 inches 106 times a minute. The pulp flowed without classification to a double-deck Deister table where some concentrate was taken off, the remaining middling product going through an injector to a single-deck Deister. Another injector sent the overflow from this table to a Union vanner for the recovery of the finer chromite. The gangue of the ore is tough and hard and the stamp capacity is less than for gold ore. Roy Stapler, in charge at the Champion, considers that the saving was about 75%, and believes that not only the percentage of $Cr_2O_3$ in the concentrate, but also the percentage of recovery depends on the percentage of iron in the ore. Ore from the Hoeft Lease and the Smith Lease have been treated at the Champion mill. The former ore is much higher in iron and lower in chromite than the latter, and the concentrate it gives contains only about 32% $Cr_2O_3$, whereas Smith's concentrate is much higher grade. The custom charge for milling, concentrating and sacking was $2.50 a gross ton.

**Nevada County Chrome Company's Plant** (Oustomah Mill). The old mill of the Oustomah mine, with some added concentrating equipment has been in steady operation since March, 1918, on chromite ore from the company's pit north of Deer Creek, near the Champion mine, and when visited in September, 1918, was going with enough ore for perhaps two months more work. Even if new ore is opened at the mine, wet weather may interfere with the three mile haul to the mill, so the future production is problematical.

The ten 1250-lb. stamps are operated by a 25-horsepower motor. They crush rather fine, and discharge to a hydraulic classifier, the effectiveness of which is open to question. The overflow (fine) from the classfied goes to a Johnson belt concentrator which gives fine concentrate. The coarse from the classifier goes to the first Overstrom Universal Concentrator. This gives some concentrate and a middling which is sent back to the stamps by injector for further crushing. The tailings from these two concentrators go to a settling box from which the pulp is lifted by bucket elevator to a second Overstrom Universal Concentrator. This gives the final concentrate. A 5-horsepower motor operates the tables and bucket elevator.

Tests on the tailing seemed to show that serious loss was occurring in the middling product, because crushing was not fine enough, rather than in the sliming, according to Morgan. For this reason, they are returning the middling for a second crushing to unlock the finely disseminated chromite. The chromite content in 40-mesh tailing was said to be much higher than in 50-mesh. This may be a special case, arising from the character of ore, as in most mills the loss has been attributed to sliming. An ore containing from 13% to 15% $Cr_2O_3$ has given after concentration a tailing carrying at times as much as 4.5% $Cr_2O_3$. The high iron content (about 10%) limits the grade of concentrate, which is said to average 36% $Cr_2O_3$. The silica content has not caused trouble. The gangue is hard and tough, resulting in a low capacity per stamp.

## COPPER.

**Arkansas Traveler Claim.** Owners, J. M. Thomas, Grass Valley; J. Hurshel, Hammonton. Under bond to J. E. Kerr, Monadnock Building, San Francisco.

> Location: Spenceville District, Sec. 13, T. 15 N., R. 6 E., 2½ miles north of Spenceville.
> Bibliography: U. S. Geol. Survey, Smartsville Folio.

The Arkansas Traveler is a patented claim of 20 acres, located on a replacement deposit of pyrite and chalcopyrite with small amounts of galena and sphalerite, carrying low values in gold and silver. The ore, associated with quartz, epidote, and calcite, occurs along a shear zone

7—46900

in altered diabase near its contact with an area of granodiorite. Very
little development work has been done in this district.

**Berkeley Prospect.** Owners, J. W. Watson and L. W. Williams,
Grass Valley.

Location: Rough and Ready District, Secs. 6 and 7, T. 16 N., R. 8 E., 6 miles
west of Grass Valley. Elevation 2000'.

This prospect is situated on a rolling plateau about 1¼ miles south of
the South Yuba River, near its tributary, Owl Creek. The property
consists of 480 acres of patented agricultural land. It was first discovered in 1904 and was abandoned after a shaft had been sunk to a
depth of 35 feet. The shaft was reopened in 1913 by L. W. Williams
and continued to a depth of 60', with a north drift 27' in length. The
ore-bearing zone occurs in an altered area of diabase schists and Calaveras slates near their contact with granodiorite; it occurs as an
impregnation and replacement of the schists by mineral bearing solutions carrying pyrite, chalcopyrite and gold. In the oxidized zone,
which extends to a depth of 50 feet, the ore is said to run $17 per ton
in gold and silver, and some native copper has been found. In the
lower levels massive chalcopyrite with secondary chalcocite occurs, and
the ore-body is said to average 3% to 4% copper. $6 in gold, and $1.50
in silver for a width of 6 feet. The only equipment on the property
consisted of a 4-horsepower gasoline hoist and pump. Some work was
done in 1914.

[Williams reported in September, 1918, that 9 cars of ore had been
shipped. It averaged 2.12% copper, 5% to 10% zinc and $5 in gold
and silver. Idle in September, 1918.]

**Big Bend Claim.** Owner, A. F. McPherson, French Corral.

Location: French Corral Mining District, Secs. 25 and 36, T. 17 N., R. 7 E.,
15 miles southeasterly by good mountain wagon road to Grass Valley.
(N. C. N. G. R. R.)
Bibliography: Cal. State Min. Bur. Bull. No. 50, page 199. U. S. Geol. Survey
Folio 18, Smartsville.

The Big Bend property comprises an area of 60 acres in locations.
The gossan is an iron-stained schist, 120' wide in places on the surface,
but the total development is an 18' adit passing through schist. The
ore carries gold, silver, and copper.

**Bitner Group.** Owner, C. C. Bitner, Spenceville, California.

Location: Spenceville Mining District, Sec. 13, T. 15 N., R. 6 E., 2 miles north
of Spenceville, thence 17 miles by road southeasterly to Wheatland (S. P.
R. R.). Elevation 900'.
Bibliography: Cal. State Min. Bur. Bull. No. 50, Copper Resources of California. U. S. Geol. Survey Folio 18, Smartsville.

The Bitner mining property consists of twelve locations, including
the Sacramento, Main Chance, First Chance, Advance Chance, and
Mother Lode claims. The abrupt slope of the surface makes tunnel

mining possible. Considerable heavy pine timber stands on the property and in the surrounding country.

About the center of the Mother Lode claim a tunnel has been started for the purpose of crosscutting the formation here exposed and to also crosscut other veins known to outcrop on the hill ahead. The tunnel is at present 450' long. The formation driven through is well mineralized for the first 300'. Near the face of this tunnel another vein was cut about 60' in width, an analysis of the ore showing $17\frac{1}{2}$% aluminum oxide and 3% magnesium. Some of this ore tested on a working basis by the Reid Electric Furnace at Newark, N. J., gave a recovery of $2\frac{1}{2}$%, or $78 per ton in magnesium. On the south side of the tunnel a shaft has been sunk to a depth of 100', and on the hill ahead of the present tunnel face, a shaft has been sunk on another vein. The latter shaft is down 65' and shows a 5' vein that assays 5-10% copper. Near the south end line of the Bitner claims and the north end line of the Mineral Hill claims, four different veins have been prospected by shaft and tunnel, and all proved to contain good values in copper and gold.

Twelve parallel veins all showing chalcopyrite are exposed on the surface, the widths varying from 5'-30' at the outcrops. The general trend of the veins is N. 26° W. with the dip easterly. The formation that incloses the mineral is a hard dark blue diabase.

Water, in any quantity desired, can be had at a reasonable rate from the Excelsior Water Company, and the property is within three miles of an electric power line.

**Boss Mine.** Owner, North San Juan Copper Company; Chas. Page, Merchants Exchange Building, San Francisco.

Location: North San Juan Mining District, Secs. 1 and 12, T. 17 N., R. 7 E., and Secs. 6 and 7, T. 17 N., R. 8 E., 1 mile west of North San Juan, thence 13 miles southeasterly by good wagon road to Nevada City. (N. C. N. G. R. R.) Elevation 2100'.
Bibliography: Cal. State Min. Bur. Bull. No. 50, page 201. U. S. Geol. Survey Folio 18, Smartsville.

The Boss Mine comprises 2 patented claims, the Aurora and the Magnetic. They were worked for gold some years ago.

The vein has been developed by a 250-foot, 2-compartment vertical shaft with a 34-foot crosscut to the vein at the 150-foot level.

The mineralization occurs in the amphibolite schist near its contact with the granodiorite as an impregnation of the schist by auriferous sulphides and the ore consists of copper, gold, silver, iron pyrites, iron oxide and chalcopyrite. The vein is 2 feet wide at the 100-foot level and 12 feet wide at the 150-foot level. It strikes north and dips 80° east. The gossan is an iron-stained schist, 1000 feet wide in places, and the oxidized zone is about 100' deep.

Equipment consists of a 50-horsepower electric motor, one 8-drill compressor, one 2-cylinder air-driven geared hoist, shaft house, power

84 MINES AND MINERAL RESOURCES.

house, blacksmith shop and 6-post head frame. Electricity is used for power.

**California Mine.** Owner, California Gold and Copper Company, Mart. Rourke, agent, Malad City, Idaho.

Location: Spenceville Mining District, Sec. 12, T. 15 N., R. 6 E., 3 miles north of Spenceville, thence 17 miles by good wagon road southwesterly to Wheatland (S. P. R. R.). Elevation 560'.
Bibliography: Cal. State Min. Bur. Bull. No. 50, page 203. U. S. Geol. Survey Folio 18, Smartsville.

The California mine covers an area of 65 acres. Development consists of a 1000-foot main adit, giving 224' of backs and drifts aggregating a total length of 600' driven on the vein.

The vein is 4' wide, strikes N. 48° W. and dips 60° NE. The ore contains malachite, bornite, chalcopyrite, iron pyrite and quartz, with a reported content of 6% copper and some gold and silver. The walls are diorite, the gossan iron-stained diorite and limonite.

The property is equipped with a blacksmith shop and office.

**California Gold and Copper Company.** Owner, California Gold and Copper Company, Mart. Rourke, agent, Malad City, Idaho.

Location: Spenceville Mining District, Sec. 13, T. 15 N., R. 6 E., 2 miles north of Spenceville, thence 17 miles by road southwesterly to Wheatland (S. P. R. R.). Elevation 600'.
Bibliography: Cal. State Min. Rept. XIII, page 60. Cal. State Min. Bur. Bull. No. 50, Copper Resources of California. U. S. Geol. Survey Folio 18, Smartsville.

The California Gold and Copper Company's property comprises five claims, the Philadelphia, American, San Francisco, Sixteen to One, and Jackson, all patented. The abruptness of the surface makes tunnel mining possible, and there is a good stand of heavy pine timber on and near the claims.

On the Sixteen to One claim, a shaft has been sunk in a 20' vein to a depth of 30', all in ore consisting of pyrite and chalcopyrite, with quartz and diabase gangue matter, and carrying from 7% to 14% of copper. On the Jackson claim the improvements consist of a 450' tunnel on the ledge, started near the south end of the claim. A body of ore, most of which was extracted and shipped, was passed through near the mouth of this tunnel. It is said to have yielded 8% of copper and $10 in gold. The vein is from 6" to 7' in width. There are croppings beyond the breast of the tunnel 60' wide. There is also a shaft 50' deep in ore. The general strike of the veins is north, and the dip is easterly. The company has been driving a long tunnel to obtain several hundred feet of backs on its claims.

Water for power is obtainable from the Excelsior Water Company, one of whose ditches passes along the top of the hill.

### Carlisle Claims.

Location: Meadow Lake Mining District, Sec. 4, T. 17 N., R. 13 E., 9 miles by good mountain wagon road southeasterly to Cisco (S. P. R. R.), 7 miles southwesterly by trail to Crystal Lake (S. P. R. R.). Elevation 5600'.
Bibliography: Cal. State Min. Bur. Bull. No. 50, page 202. U. S. Geol. Survey Folio 39, Truckee.

The Carlisle property comprises an area of 260 acres of patented ground.

The vein is developed by adits, totaling 1300' in length; No. 1 adit is 500' long; No. 2 is 400' long, and the others total 400'.

The vein strikes NW.-SE. and varies in width from 2' to 10'. The filling is quartz, containing pyrite and chalcopyrite, and the walls are porphyry. The gossan is an iron-stained porphyry. Reported ore values are copper $7\frac{1}{2}\%$, gold $6 and silver 4 oz.

The property is equipped with a 2600-foot aerial tramway, a 200-horsepower turbine wheel, sawmill, blacksmith shop, boarding house and bunk house. Water is used for power.

**English Mountain Copper Mine** is located in Sec. 6, T. 18 N., R. 13 E., 13 miles east of Graniteville, and is bonded to Henry Schroeder of Nevada City. The old workings of the English Mountain Gold Mine adjoin. Schroeder reports that a copper bearing gossan has been traced for a width of 50 feet, and a length of 1000 feet; alongside this is a broken foot-wall of the diabase country rock, carrying chalcopyrite stringers, and widely varying amounts of gold. About 1000 feet of drifts and crosscuts have been driven, one 500-foot drift along the strike, and another 300 feet long, above it, being the principal work done. The following assays are said to have been made by the University of Nevada:

|  | Copper. per cent | Iron. per cent | Sulphur. per cent |
|---|---|---|---|
| Pyrrhotite-pyrite | 5.0 | 45.9 | 35.4 |
| Bornite-pyrite | 2.4 | 53.5 | 27.7 |
| Pyrrhotite-Marcasite | 1.3 | 51.4 | 28.3 |

The property has been examined by W. W. Adams and others, for the U. S. Smelting and Refining Company; by Carlton Bray for the International S. and R. Company, and by other engineers.

The property is in a high and rugged country, well over 6000 feet in elevation, and the winter climate is rigorous. The country is ordinarily accessible about one-half of the year. The adjoining English Mountain Gold mine, which has been credited with $100,000 production, has not been actively worked since the surface plant burned in 1897. The best gold ore in these claims is said to occur at the crossings of andesite dikes and the copper-bearing zones.

**Eureka Consolidated Claims.** Owner, Eureka Consolidated Mining Company; Mary E. Mooney, agent, Smartsville, California.

Location: Spenceville Mining District, Secs. 1 and 12, T. 15 N., R. 6 E., 3 miles north of Spenceville, thence 17 miles southwesterly by road to Wheatland (S. P. R. R.). Elevation 600'.
Bibliography: U. S. Geol. Survey Folio 18, Smartsville.

**Fairview.** (See under Gold.)

**Gautier Ranch.** Owner, Wm. Gautier, Auburn, California.

Location: Spenceville Mining District, Secs. 28 and 29, T. 14 N., R. 8 E., 9 miles southerly by good wagon road to Auburn. Elevation 1450'.
Bibliography: Cal. State Min. Bur. Bull. No. 50, page 203. U. S. Geol. Survey Folio 18, Smartsville.

The Gautier Ranch property covers a total area of 320 acres.

The vein developed by an old vertical shaft, 80' deep, is filled with quartz porphyry, carrying malachite, chalcopyrite and pyrite. It strikes N. 4° W. and dips 80° east. The walls are schist. The gossan is an iron-stained porphyry and rhyolite with some limonite, thirty feet wide in places.

**Grizzly Ridge.** (See under Gold.)

**Last Chance Mine.** Owners, Miller & Roberts, Spenceville, Nevada County, California.

Location: Spenceville Mining District, Secs. 12 and 13, T. 15 N., R. 6 E., 2½ miles north of Spenceville, thence 17 miles southwesterly by road to Wheatland (S. P. R. R.). Elevation 900'.
Bibliography: Cal. State Min. Bur. Bull. No. 50, Copper Resources of California, page 195. U. S. Geol. Survey Folio 18, Smartsville.

The Last Chance property comprises one patented claim of that name. Considerable heavy pine timber stands on the claim, and covers the adjoining country. Tunnel mining is possible, due to abruptness of the hills.

The Last Chance is opened by a shaft, 240' deep, equipped with a steam hoist. The vein is 5' to 8' wide, carrying pyrite and chalcopyrite in considerable quantities. Ore was shipped from the mine as early as 1876. The pyrite occurs in this mine as well as in all others in the district, in lens-shaped bodies, the inclosing formation being a hard, dark-blue diabase. The general strike of the vein is N. 26° W. and it has an easterly dip.

Water power is available from the Excelsior Water Company, one of whose ditches passes along the top of the hill.

**Little Gem.** (See under Gold.)

**Lotzen Ranch.** Owner, Wm. Lotzen, Grass Valley.

Location: Spenceville Mining District, Sec. 2, T. 14 N., R. 7 E., 10 miles southwest of Grass Valley (N. C. N. G. R. R.) by good wagon road. Elevation 1050'.
Bibliography: Cal. State Min. Bur. Bull. No. 50, page 204. U. S. Geol. Survey Folio 18, Smartsville.

The Lotzen Ranch property covers an area of 158 acres. There is a 60-foot adit crosscut on the property, 16 feet of which is in a mineralized diorite, showing chalcopyrite, bornite, pyrite, quartz and red oxide of copper. This appears to be near the water level. The footwall is granodiorite and the hanging wall, diorite. The vein which is 16 feet in width, strikes N. 30° W. and dips 85° southwest, with a gossan of iron-stained diorite and limonite.

**Mammoth Gold Copper Mine.** Owner, G. W. Broyles, French Corral.

Location: North San Juan Mining District, Sec. 12, T. 17 N., R. 7 E., 2½ miles southwest of North San Juan, thence 13 miles southeast to Nevada City (N. C. N. G. R. R.) by good wagon road.
Bibliography: Cal. State Min. Bur. Bull. No. 50, page 204. U. S. Geol. Survey Folio 18, Smartsville.

The Mammoth Gold Copper mine, comprises locations covering an area of 60 acres. The only development work, a 40-foot adit near the level of the Yuba River, exposes a vein dipping 85° east and striking north between walls of slate. The ore has a reported value of 2% copper and $3 gold.

**Mammoth Group.** (See under Gold.)

**Mineral Hill.** Owner, Mineral Mining and Smelting Company; H. Gimbel, agent, 4051 Seventeenth street, San Francisco.

Location: Spenceville Mining District, Sec. 13, T. 15 N., R. 6 E., 2½ miles north of Spenceville, thence 17 miles by road southwesterly to Wheatland (S. P. R. R.). Elevation 750'.
Bibliography: Cal. State Min. Bur. Rept. XII, page 67; XIII, page 60, Cal. State Min. Bur. Bull. No. 50, Copper Resources of California, page 195. U. S. Geol. Survey Folio 18, Smartsville.

The Mineral Hill property embraces the Golden Gate, Legion and Index patented claims. The abruptness of the hills makes tunnel mining possible. An abundance of heavy pine timber stands on the claims and covers the adjoining country.

The Golden Gate is opened by a shaft 15' deep, showing a 4' vein of copper and iron sulphides. On the north line, on a vein west of and parallel with the Last Chance vein, is an inclined shaft 150' deep. To the east of these veins there is a tunnel 70' long driven 300' on another vein, and at its end is a west crosscut to the ledge. A deep tunnel is being driven to obtain 500' of depth under the best of the claims. This tunnel has cut three very promising ledges.

Thirteen parallel veins outcrop on the property, varying in width from 5' to 30' at the surface. All show chalcopyrite. A considerable amount of carbonate ore was shipped from the Golden Gate, and is said to have had a value of $12.50 gold per ton and 10% copper. All ore shoots pitch to the north, while the general strike of the veins is N. 26° W. with an easterly dip.

Water for power could be obtained from the Excelsior Water Company, one of whose ditches passes along the top of the hill.

**Mother Lode Group.** (See under Gold.)

**Mountain Lion.** (See under Gold.)

**Native Son.** (See under Gold.)

**Oro Grande Group.** Owners, J. P. Clark and F. J. Cook, Cisco.

Location: Meadow Lake Mining District, Sec. 27, T. 18 N., R. 13 E., 9 miles north of Cisco (S. P. R. R.), by fair mountain wagon road. Elevation 7350'.

The Oro Grande group of locations comprise a total area of 80 acres.

This property was worked fifty years ago, until the sulphide zone was reached, as a free milling quartz mine. One man is now doing assessment work.

The property has been developed by 3 shafts. No. 1 is a vertical shaft 32' in depth; No. 2 is a 3-compartment vertical shaft 125' deep, now partly filled; No. 3 is a 2-compartment vertical shaft, 105' deep, now caved. There are 200' of open cuts and a 30' adit on the vein which varies from 8'–50' in width, strikes N. 85° W. and dips 72° south. The walls are porphyry, the ore being chalcopyrite, bornite, malachite, pyrite and quartz. The gossan is an iron-stained porphyry. The reported value of the ore is copper 5%, gold $7, and silver 3 ozs. There are 20 parallel veins exposed in places, with croppings for 4500'. Oxides are from 1'–10' in depth. A 4-stamp mill stands on the property.

**Peter.** (See under Gold.)

**Pine Hill Mining Company.** Owner, Pine Hill Mining Company, Wolf post office; J. A. Robles, superintendent.

Location: Spenceville Mining District, Sec. 12, T. 14 N., R. 7 E., 13 miles northeast of Auburn (S. P. R. R.) by good wagon road. Elevation 1750'. Bibliography: Cal. State Min. Bur. Bull. No. 50, page 199. U. S. Geol. Survey Folio 18, Smartsville.

The Pine Hill Mining Company's property comprises a total of 160 acres, embracing the three patented claims, Golden Gate, Golden Gate Extension and Thrasher. The property has been worked as a gold mine.

Development consists of a vertical 2-compartment main shaft 200' deep, and other shafts and inclines aggregating 200'. Drifts totaling 1100' have been run.

The deposit contains gold, native copper, silver, chalcopyrite, malachite, red oxide, bornite, clay, quartz, pyrite and iron oxide. Fine white kaolin, 150' wide and some barite occur on the property. The footwall is diabase, the hanging wall serpentine. The gossan is iron-stained porphyry and rhyolite, limonite and hematite. Vein matter is 27' in width, strikes N. 3° W., and dips 62° east. Water from the adit running through sluices precipitates some cement copper and iron.

Equipment comprises a 10-stamp mill, 1000-lb. stamps, 3 vanners, amalgamating plates and full equipment for free-milling ore, blacksmith shop, 2-drill compressor, 8-horsepower geared hoist, boarding house, bunkhouse, mill building, assay office, office building and 2 pumps. Steam is used for power, 85-horsepower being developed.

**Plumas Jumbo and Little Jumbo Mines.** (See under Gold.)

**Queen Regent Mine.** Owner, Queen Regent Merger Mining Company, San Francisco.

> Location: Spenceville Mining District; Spenceville is 17 miles by road, northeast of Wheatland (S. P. R. R.).
> Bibliography: U. S. Geol. Survey Folio 18, Smartsville.

It was reported early in 1916 that the Mineral Hill Mining District was experiencing a revival as the result of work done during the past two years by the Queen Regent Company. The company contracted for a complete treatment plant, which was finished during March, 1916. This was expected to treat the ore at a cost of $3.50 per ton. A percentage of profit applies on the purchase price until the plant is paid for in full, after which it will be taken over and operated by the company. The engineer for the parties interested in the deal reports sufficient high grade ore developed to pay for the plant in 90 days as the ore assays from $12 to $20 per ton, some reaching $50 in gold, silver and copper. A large amount of new underground work is being undertaken and indications are good for a steady output.

The plant is designed to handle the high-grade ores of the Woehler mine, containing copper and gold. Concentrates were shipped for smelting, but later a smelting plant of 15 tons daily capacity was erected.

**Red Ledge Mine.** Owners, N. F. McPherson et al., French Corral.

> Location: French Corral Mining District, Sec. 36, T. 17 N., R. 7 E., 15 miles southeasterly to Grass Valley (N. C. N. G. R. R.), by good mountain wagon road. Elevation 1220'.
> Bibliography: Cal. State Min. Bur. Bull. No. 50, page 200. U. S. Geol. Survey Folio 18, Smartsville.

The Red Ledge mine comprises a total area of 160 acres made up of locations. For some years the property was worked as a gold mine.

It has been developed by six adits. No. 3 adit is 110 feet in length, and opens up 7 feet of ore. No. 2 adit is 167 feet in length, with 27 feet of ore and No. 6 adit is 330 feet in length, with 16 feet of ore. Adits Nos. 1 and 4 are old workings and are caved. Nos. 2 and 3 are about 100' above No. 6.

The ore is made up of gold, silver, copper, malachite, chalcopyrite, iron pyrite and quartz, with schist walls. The vein dips 60° west and strikes N. 45° W. The gossan is an iron-stained schist, in places 500' wide.

Equipment consists of a 10-stamp mill with 1000-lb. stamps, water power impulse wheel, and cyanide plant.

**South Yuba Claim.**  (See under Gold Lode.)

The ore here carries 1%–3% copper. There has been no recent activity. Some years ago it was planned to ship the ore for the sulphur content, but this idea was evidently not carried out.

**Spence Mineral Company.**  Owner, Spence Mineral Company, 528 Mills Building, San Francisco.

> Location: Spenceville Mining District, Secs. 25 and 27, T. 15 N., R. 6 E., at Spenceville.  Elevation 500′.
> Bibliography: U. S. Geol. Survey Folio 18, Smartsville.

The property embraces two patented claims, known as the San Francisco and Grass Valley and two millsites of the same names respectively. The company also owns some adjacent patented agricultural ground.

This copper deposit was discovered in 1862 and in the following 18 years a shaft was put down to a depth of 180 feet and the ore-body prospected by levels from the 60′, 100′ and 150′ stations. A small smelting plant was also erected. In 1880 a cave-in occurred which destroyed the shaft and plant and deep exploration has never since been attempted, although the mine has been worked at intervals as an open cut. No work was done from 1906 to 1914. Operations were to have been started in 1915, but the entire surface plant was destroyed by fire October 30, 1915.

The deposit occurs along a fissured and altered zone in diabase near its contact with granodiorite, and the primary sulphides pyrite and chalcopyrite, occur as an impregnation and as a local enlargement of the quartz veins which traverse the zone. The mineralized zone has at places a width of 100 feet. It is capped by iron gossan, below which an horizon of secondary enrichment was encountered and ore (bornite) carrying as high as 20% copper is said to have been mined from the open cut. Undoubtedly in depth the pyrite and chalcopyrite will contain only a small percentage of copper. The pyrite has a maximum width of from 30′ to 50′ of solid sulphide. About the year 1900, 15 tons of pyrite per day were being shipped to San Francisco for the manufacture of sulphuric acid.

**Sweet Ranch.**  (See under Gold.)

**Tola Group.**  Owner, Tola Mining Company; A. R. Tabor, agent, Auburn.

> Location: Meadow Lake Mining District, Sec. 27, T. 18 N., R. 13 E., 8 miles south to Cisco (S. P. R. R.) by fair mountain road.  Elevation 7150′.
> Bibliography: Cal. State Min. Bur. Bull. No. 50, page 201.  U. S. Geol. Survey Folio 37, Truckee.

The Tola Group comprises three patented claims: the Alice, Bulah and Sunnyside, with a total area of 60 acres.

The only development work done on the claims is a 50' open cut, a 12' adit and three 10' holes. The vein is 4' in width, dips 49° south and strikes east between walls of porphyry. The ore contains chalcopyrite, pyrite and malachite, and values reported on 15 to 18 tons shipped in 1906 averaged $28 per ton in gold, silver and copper. The gossan is an iron-stained porphyry.

Equipment consists of a bunk house, boarding house and barn.

**Turner Group.** Owner, W. M. Turner, Webber Lake, Sierra County.

Location: Meadow Lake Mining District, Sec. 16, T. 18 N., R. 13 E., 10 miles north of Cisco (S. P. R. R.) by fair wagon road. Elevation 7000'.
Bibliography: Cal. State Min. Bur. Bull. No. 50, page 201. U. S. Geol. Survey Folio 37, Truckee.

The Turner group of locations cover an area of 120 acres.

The property has been developed by some surface cuts and an 18-foot shaft, exposing a vein 16' in width at the shaft, striking northwesterly and dipping to the south. Its walls are porphyry, and the ore contains chalcopyrite and pyrite with a value reported as gold $9, silver 5 ozs. and copper 4%. The gossan is an iron-stained porphyry.

**Wetteran Ranch.** Owner, G. G. Wetteran, R.F.D. box 62, Grass Valley.

Location: Spenceville Mining District, Sec. 34, T. 15 N., R. 7 E., 4½ miles east of Spenceville; about 10 miles by good wagon road southwest of Grass Valley (N. C. N. G. R. R.). Elevation 900'.
Bibliography: Cal. State Min. Bur. Bull. No. 50, page 201. U. S. Geol. Survey Folio 18, Smartsville.

The Wetteran Ranch property covers a total area of 540 acres.

Development consists of a 20-foot vertical shaft, open cuts and trenches. The vein strikes to the north, the filling containing malachite and pyrite. The gossan is limonite and iron-stained breccia of rhyolite.

### GOLD—DRIFT MINES.

**Albert and Yosemite Mine.** Owners, Albert Gravel Mining Company; W. D. Harris, Grass Valley; V. M. Thomas, Grass Valley.

Location: Nevada City District, Secs. 31 and 32, T. 17 N., R. 9 E., 2 miles northeast of Nevada City.
Bibliography: U. S. Geol. Survey Folio 29, "Nevada City Special."

This property, with a total area of about 160 acres, was last worked as a drift mine about 1894, when rich gravel is said to have been taken out. The channel worked is a branch of the Harmony Channel and is covered by a capping of rhyolite and andesite from 300 to 500 ft. in thickness. The gravel was worked through the Yosemite incline. No information is available as to the amount of gravel breasted, production, or area of gravel remaining to be worked.

**Alta-California Mine.** Owner, L. Haas Estate, Grass Valley.

Location: Rough and Ready Mining District, Secs. 19 and 30, T. 16 N., R. 8 E.,
2½ miles west of Grass Valley. Elevation 2000'.
Bibliography: U. S. Geol. Survey Folio 18, Smartsville.

This property of 160 acres covers about 4000 feet along a branch of
the Alta Hill-Rough and Ready channel. No work has been done in
recent years.

**Alta Hill Mine.** Owners, W. J. and T. W. Mitchell, Grass Valley.

Location: Grass Valley District, Sec. 22, T. 16 N., R. 8 E., 1 mile north of
Grass Valley. Elevation 2600'.
Bibliography: U. S. Geol. Survey, Nevada City-Grass Valley Special Folio.
U. S. Geol. Survey Prof. Paper No. 73.

The Alta Hill mine was last worked about 25 years ago, through the
Godfrey's vertical shaft which is 225' in depth. Some drifting was
done on a channel in a fine cemented quartz. This property has a total
area of 112 acres adjoining the old Alta No. 1 and No. 2 claims, in which
the Alta channel was worked for a length of 3000' and a width from
50' to 150'. The production of this area is said to have been $1,000,000.

**Bigg and Sims Group.** Owners, T. H. McGuire et al., Grass Valley.

Location: Grass Valley District, Sec. 1, T. 15 N., and Secs. 35 and 36, T. 16 N.,
R. 8 E., 1½ miles southeast of Grass Valley. Elevation 2700'.

This property consists of the Bigg and Sims patented placer claims,
and mineral rights under 160 acres of agricultural land, the surface
rights to which are owned by Ducoty and Gill. The 360 acres controlled is supposed to contain about ¾ of a mile of channel capped with
from 20 to 80' of andesite. The general course of the channel is north
and the deposit, from 4 to 15' deep, is composed of andesite and diabase
gravel with rough quartz showing free gold. The character of the
quartz boulders is said to be entirely different from that of the Empire
vein system lying to the west. The channel has an average width of
50' and a depth of 3 feet. A tunnel was driven in the Ducoty claim
for a distance of 400' in the channel and the ground thus developed
was breasted. This work was done in 1894, and since that time the mine
has been practically idle.

**Blue Lead Claim.** Formerly owned by A. Roach and C. Haggerty
of Moons Flat.

Location: Relief Hill District, Secs. 3 and 4, T. 17 N., and Secs. 33 and 34,
T. 18 N., R. 10 E. Elevation (tunnel) 3649'.

The Relief Hill channel was supposed to pass through this property
and a 2500' tunnel was driven from Logan Ravine in a northwesterly
direction. Some gravel was breasted from the tunnel, but it is doubtful
if it came from the main channel.

**Blue Tent Mine.** Owner, Eleanor Hoeft, Nevada City.

Location: Nevada City Mining District, Sec. ?, T. 17 N., R. 10 E., about 5 miles northeast of Nevada City.
Bibliography: U. S. Geol. Survey, W. Lindgren, Prof. Paper 73, pages 125–132. U. S. Geol. Survey, W. Lindgren, 17th Annual Report, pt. II, pages 1–262, 1896. U. S. Geol. Survey Folios 18 and 29, Smartsville and Nevada City.

This old gravel mine was reopened in December, 1914, by D. A. Campbell and others. The tunnel was cleaned, and drifting toward the channel was in progress. It was formerly a large producer.

In 1917 some hydraulic mining was done on this ground and work planned resumed in the winter of 1918. The mill for crushing cement gravel was burned during the summer of 1918.

**Cherokee Mine.** (See River Mines Company.)

In January, 1915, the old 400' tunnel had been cleaned out, and was being advanced 400' to connect with an old shaft sunk in early days when the mine was a producer.

**Cold Springs Mine.** Owners, Cold Springs Quartz and Gravel Mining Company; T. N. Coan, Nevada City; W. G. Motley.

Location: Nevada City District, Sec. 4, T. 16 N., and Sec. 33, T. 17 N., R. 9 E.
Bibliography: U. S. Geol. Survey Folio 29, Nevada City Special.

The Cold Springs property consists of about 450 acres, covering approximately 5000' along the eastern extension of the Harmony channel, which has been successfully worked by drifting from the Cold Springs tunnel and incline and from the East and West Harmony inclines on adjoining ground. The Cold Springs tunnel, 1500' in length, and the incline, 700' in length, cut the channel about 1000' apart and the ground between has been profitably mined. The channel, which is from 80' to 150' in width at this point, is covered with a capping of rhyolite and andesite to a depth of 450'. The gravel in this channel is for the most part cemented and it must therefore be crushed in a stamp mill. The Buckeye quartz mine is also owned by this Company. See Buckeye.

**Delaware Mine** (West Harmony). Owners, C. H. Mallen et al., Nevada City.

Location: Nevada City District, Sec. 5, T. 16 N., and Sec. 32, T. 17 N., R. 9 E., 2 miles northeast of Nevada City. Elevation 3000'.
Bibliography: Cal. State Min. Bur. Repts. XI, page 300; XII, page 202; XIII, page 240. U. S. Geol. Survey Folio 29, Banner Hill Special.

The Harmony Channel, which is from 150 to 250' in width, has been extensively worked in this property and in the Harmony placer mine, adjoining it on the east. The property consists of 160 acres of patented ground. The channel is covered here with 600' of rhyolite and andesite, and has been worked through the West Harmony incline, which is 520' long and a drift 1100' in length. Most of the channel that was accessible through this opening was worked out during the operations which began in 1892. The gravel was partially cemented and was crushed in a 15-stamp mill.

**Derbec Group.** Owner, Derbec Blue Gravel Gold Mining Company; B. E. Anger, 1226 Waller Street, San Francisco.

> Location: North Bloomfield District, Secs. 29, 30 and 31, T. 18 N., R. 10 E., 1 mile north of Bloomfield. Elevation 3800'.
> Bibliography: Cal. State Min. Repts. VIII, page 456; XI, page 311; XIII, page 241. U. S. Geol. Survey Prof. Paper 73, Tertiary Gravels of the Sierra Nevada (Lindgren). U. S. Geol. Survey Folio 66, Colfax.

This property, consisting of the Montreal, Skidmore, Brockmeyer, Ultimatum and parts of the Franklin and Last Chance claims, comprises an area of 620 acres. The mine was successfully worked from 1877 until 1893, when the main channel had been practically worked out. It is credited with a production of $2,000,000, the gravel averaging $2.47 per ton. The channel was worked for a distance of 7000' east of the shaft, varying in width from 150 to 600' and was breasted from 8 to 16' above bedrock. The Union Blue Gravel Mining Company controls the eastern extension of the Derbec channel.

**Eagle Echo Claim.** Owner, Eureka Lake and Yuba Canal Company; Geo. W. Starr, Grass Valley.

> Location: Relief Hill District, Secs 4 and 9, T. 17 N., R. 10 E., 3 miles east of North Bloomfield. Elevation 3700'.
> Bibliography: Cal. State Min. Bur. Rept. XIII, page 241. U. S. Geol. Survey Folio 66, Colfax. U. S. Geol. Survey Prof. Paper 73, Tertiary Gravels of Sierra Nevada (Lindgren), pages 140–141.

**El Oro (Ballarat) Claim.** Under option to El Oro Consolidated Gravel Mining Company; F. R. Phelps, North Bloomfield, Nevada City.

> Location: North Bloomfield District near Lake City.
> Bibliography: U. S. Geol. Survey Folio 66, Colfax.

In December, 1914, this company started sinking a 2-compartment shaft to develop a buried channel. No details could be obtained, but it is believed that the gravel channel will be encountered at a depth of 200'.

Arrangements were being made in March, 1916, for the installation of a hoisting and pumping plant, and the construction of the headframe was progressing.

**Fong Wing Group.** Fong Wing was lessee of the You Bet Mining Company's property, in part.

> Location: You Bet Mining District, Secs. 26, 28, 30, 31, 25, 36, 32, Twps. 15, 16 N., Ranges 10, 9, E., 7 miles from Dutch Flat. Elevation 2600' to 3200'.
> Bibliography: Cal. State Min. Bur. Rept. XIII, page 236. U. S. Geol. Survey, W. Lindgren, Paper 73, page 144. U. S. Geol. Survey Folio 66, Colfax.

The claims which were leased to Fong Wing are Birdseye Canyon Placer, You Bet, Neice and West, Browns Hill, Walloupa, Washington, Brown Bros., Poverty, Atkins & Taylor, Rose Duryea, Gail, Emigrant, Smith-Powell, Chicken Point, California Nos. 1, 4, 6, 7 and 8, Missouri Canyon, and Newark Fluming and Milling Company.

Most of these claims are included in the property of the You Bet Mining Company, which started hydraulic mining operations in the winter of 1913–1914. They worked successfully until August, 1914, when restrained by a court injunction. Fong Wing leased the above claims on August 25, 1915. The property is said to have produced in the neighborhood of $3,000,000 previous to 1913.

The gravel has been developed by the Nevada bedrock tunnel, which connects with the Red Dog incline by an upraise. The Red Dog incline is about 200' on the slope and has a total length of from 800' to 900'. Wing started a new tunnel near the old 'tie' workings at You Bet. There were 35 men at work in the Red Dog tunnel and 35 men shoveling tailings in Wilcox Canyon and at You Bet in 1916. The tailings are said to average $1 per yard. No timbering is required on the tunnels and hand labor was the only method of working.

The gravel of the deep channel in the vicinity of You Bet is coarse and partly cemented in the deep trough, but otherwise consists of fine quartz gravel mixed with sand. The gravels in the deep channel are from 75' to 200' in depth, and pay gravel is from 300' to 400' in width. There is said to be about three-fourths of a mile of virgin ground along "Blue Lead." (For prospective development in 1918 on the You Bet properties, see under Hydraulic Mining.)

**Fountain Head Group.** Owner, Fountain Head Gold Mining Company, Nevada City: Dr. C. L. Muller, Nevada City.

> Location: Nevada City, Banner District, Secs. 3 and 4, T. 16 N., and Secs. 33 and 34, T. 17 N., R. 9 E. Elevation 3300'.
> Bibliography: Cal. State Min. Bur. Rept. XIII, page 244.

The property of the Fountain Head Gold Mining Company contains about 375 acres of patented ground lying east of the Gold Spring Mine and it is supposed to contain the eastern extension of the Harmony channel which was worked successfully in the Harmony and Gold Spring mines. The property has not been worked for many years. There is also a small vein on this property which strikes north and dips 70° east.

**Grizzly Ridge Mine.** Owners, Mark E. Alling et al., Downieville. Under option to R. H. Franklin of Los Angeles, California.

> Location: Grizzly Ridge Mining District, Secs. 22, 23, and 24, T. 18 N., R. 9 E., 17 miles south to Nevada City.
> Bibliography: U. S. Geol. Survey Folio 66, Colfax.

This mine was worked as an hydraulic mine from 1886 until the suspension of hydraulic mining in 1894. The property remained idle until recently reopened as a drift mine by M. E. Alling and his associates. The claims controlled by this company are the Outlet, Mazuma, Grey Eagle, Polar Bear and Little Bear, an area of about 600 acres.

The ancient channel which has been proven to traverse the property for a distance of about two miles, is supposed to be the southern continuation of the famous Blue Lead Neocene channel, according to Mr. Alling who has made a special study of the ancient channels of Sierra and Nevada Counties. Owing to the erosion of various portions of the Blue Lead, Grizzly Ridge and the renowned North Bloomfield channels and to their lava capping the relation and continuity of the Neocene streams is obscure and further exploration and study will be necessary before the problems can be solved.

The general direction of the channel is westerly. The capping is andesite and rhyolite, the bedrock slate. The width of the quartz gravel is said to be from 200' to 400' and the depth from 2' to 10'. It has been developed by means of a bedrock tunnel 850' in length, costing on an average $8 per foot. Since the channel was recently encountered, drifts have been driven in the river gravel for over 400 feet. The pay gravel so far developed is 65' in width and from 3' to 6' in depth. The full width of the channel has not as yet been accurately determined. The only gravel breasted was taken from near the rim, 400' from where the new pay streak has been encountered. This gravel is reported to have yielded $1.10 'per load.' The gold is 885 fine and as a rule is small, with an occasional nugget.

The bedrock tunnel is 86' below the lowest point in the channel so far opened and in October, 1914, the channel was being opened by raises and the gravel was being blocked out by drifts and gangways. The property is equipped with four houses, blacksmith shop, stable and gravel washing equipment.

**Grover Claim.** Owner, Grover Murphy Development Company; in care W. Farris, agent, Nevada City.

> Location: Nevada City Mining District, Secs. 1 and 12, T. 16 N., R. 8 E.,
> 1½ miles northwest of Nevada City. Elevation 2650'.
> Bibliography: Cal. State Min. Bur. Rept. XIII, page 255. U. S. Geol. Survey
> Folio 29, Nevada City Special.

A 1000' tunnel was driven on the Cement Hill channel near the point where it emerges under the rhyolite-andesite capping. There is also a quartz vein on the property, but it is only superficially developed. No work had been done for many years.

**Harmony Mine.** Owner, Harmony Gold Mining Company; H. W. Brand, Nevada City.

> Location: Nevada City District, Sec. 5, T. 16 N., and Sec. 32, T. 17 N., R. 9 E.,
> 2½ miles northeast of Nevada City. Elevation 3000'.
> Bibliography: Cal. State Min. Bur. Repts. XI, page 298; XII, page 190; XIII,
> page 247. U. S. Geol. Survey, Folio 29, Banner Special. U. S. Geol. Survey
> Prof. Paper 73, Tertiary Gravels of the Sierra Nevada.

The Harmony ground lies between the Cold Spring property on the east and the Delaware or West Harmony. The main Harmony Channel branches 800' east of the boundary between the West Harmony and

Harmony claims, one branch turning to the north for a distance of 2000', where it emerges from beneath the lava capping; the other branch continues eastward into the Cold Springs ground where it was worked with considerable success.

The north channel was worked in the early days of drift mining by a tunnel from the 'blow out,' and by the Harmony shaft. Later, the Harmony incline and drift were driven through from which the main Harmony Channel and Cold Spring branch were worked. The main channel was from 100' to 200' in width and the gravel was cemented. The property was extensively operated from 1892 to 1900, and the greater part of the channel controlled by this company has been worked out.

**Jenny Lind Mine.** Owner, Jenny Lind Consolidated Mining Company; F. J. Thomas, agent, Grass Valley.

Location: Rough and Ready Mining District, Sec. 20, T. 16 N., R. 8 E.
Bibliography: U. S. Geol. Survey Folio 18, Smartsville.

A bedrock tunnel was started on this property forty or fifty years ago to develop the westerly extension of the Alta Hill channel. This tunnel was continued in bedrock to a total length of 4000', a number of raises were put up and an upper tunnel was run in the gravel encountered. The gravel, however, only averaged $1 per ton, and for this reason work was finally abandoned.

**Kate Hayes.** (See River Mines Company, Hydraulic.)

**Lupine Mine.** Owners, F. T. Nilon et al., Nevada City, California.

Location: Nevada City Mining District, Sec. 21, T. 17 N., R. 10 E., 8 miles northeast of Nevada City. Elevation 4000'.
Bibliography: Cal. State Min. Bur. Repts. XII, page 191; XIII, page 250. U. S. Geol. Survey Folio 66, Colfax. U. S. Geol. Survey Prof. Paper 73.

This property is situated on Washington ridge midway between Nevada City and Washington. A 750' bedrock tunnel was driven to develop the Quaker-Hill-Galbraith channel which runs north and south under the lava capping. The drifting operations carried on by the Lupine Company developed a supposedly tributary channel which was filled with angular wash and some coarse gold, but no large body of pay-gravel was encountered. On the south side of the ridge on the North Fork of Deer Creek, some hydraulicking was done on the channel.

The property has been idle for a number of years.

**Major Mine** (Richland or Ragon Flat Mine). Owner, Major Gold Mining Company, Santa Monica, California. Bonded.

Location: Nevada City Mining District, Sec. 2, T. 16 N., R. 8 E., 2½ miles west of Nevada City. Elevation 2740'.
Bibliography: Cal. State Min. Bur. Rept. XIII, page 260. U. S. Geol. Survey Folio 29, Nevada City Special. U. S. Geol. Survey Prof. Paper 73, pages 128–132.

This company has under bond about 7000' along the Cement Hill channel. In early days a small portion of this channel was worked

8—46900

through the shaft and rich gravel was taken from a paystreak 50' in width. The Ragon incline, through which the work at present is being carried on, lies about 500' east of the Empire shaft. The incline was put down at an angle of 45° north and is 145' in depth. From the bottom a drift has been driven 130' northwest and 60' southeast, and short branch drifts have been run across the channel from the main drift.

The width of the pay gravel has not as yet been determined, but some of the gravel encountered is said by Mr. Graham to average from $3 to $4 per cu. yd. and the depth to be 4 feet. The gravel is in part cemented, and it is the intention of the company to install a ten-stamp mill. The property at present is equipped with a 22 horsepower electric hoist and a 6" electric driven Cornish pump.

This is the only company operating on this channel at the present time.

**Manzanita Mine.**  Owner, Fred Ayer; W. F. Duboise, agent, Nevada City.

Location: Nevada City Mining District, Secs. 6 and 7, T. 16 N., R. 9 E., 1 mile
  north of Nevada City. Elevation 2800'.
Bibliography: Cal. State Min. Bur. Repts. VIII, page 458; XI, page 302; XII,
  page 191; XIII, page 251.  U. S. Geol. Survey Folios 18 and 28.  U. S. Geol.
  Survey Prof. Paper 73.

The Manzanita channel was worked in early days leaving the hydraulic pit as one of the prominent features of topography north of Nevada City. After the suspension of hydraulic mining, the rich gravel lying on the bedrock of the channel was worked by drifting from Nevada City under the lava-capped ridge to the northern exposure of the channel in Howe cut, a distance of 3000 feet. The central portion of the channel under the deep lava cap was worked by means of the Odin and Nebraska inclines in the Odin property, which adjoins the Manzanita on the north. This channel is reported to have produced over $3,000,000. The Harmony channel, which has been extensively worked in the East and West Harmony mines, about a mile northeast of Nevada City, was formerly supposed to be connected with the Manzanita channel under the lava-covered Harmony Ridge. Now the data obtainable seems to point to the conclusion that the Manzanita and Harmony channels both flowed in a northerly direction and their point of juncture was one or two miles north of Howe cut. This portion of the ancient streams has been eroded.

**Murphy.**  (See Grover.)

**Sazarack Claims.**  Owner, E. A. Roberts Estate; W. H. Lyons, agent, Stockton, California.

Location: Rough and Ready Mining District, Sec. 19, T. 16 N., R. 8 E., 3 miles
  west of Grass Valley. Elevation 2000'.
Bibliography: U. S. Geol. Survey Folio 18, Smartsville.

This property consists of the Sazarack and Bunker Hill patented placer claims, an area of about 120 acres. These claims are supposed to contain the western extension or a branch of the Alta Hill-Town Talk Channel. The bedrock is amphibolite schist, in part covered with a volcanic capping of andesite. No work has been done on the claims for a number of years.

**Sharpe Quartz and Gravel Mining Company** (Black Bear). (See under Lode.)

**Union Mine.** Owners, C. D. Jepsen et al., Relief Hill.

Location: North Bloomfield Mining District, Secs. 4 and 9, T. 17 N., R. 10 E., 3½ miles east of North Bloomfield, thence 14 miles by road southwest to Nevada City. Elevation 3600′.
Bibliography: Cal. State Min. Bur. Repts. XII, page 201; XIII, page 265. U. S. Geol. Survey, W. Lindgren, Prof. Paper 73, pages 139-141. U. S. Geol. Survey Folio 66, Colfax.

The Union mine covers gravels of the Dubec channel. At Relief erosion has exposed a deep trough in the old bedrock and about 200 acres of auriferous gravels. The oldest gravel, as usual coarser and containing less quartz, is 60′ deep and covered by 100′ to 200′ of alternating quartz sand and clay.

The Union tunnel, about 2500′ long, has been driven from the southwest side of the gravel area, and amounts up to $30,000 and $40,000 were produced for a number of years.

In December, 1915, the tunnel was being driven to intersect the bottom of an old 40′ incline shaft that was sunk to the gravel channel.

**West Harmony Claim.** (See Delaware.) Owners, C. H. Mallen et al., Nevada City.

Location: Nevada City District, Sec. 5, T. 16 N., and Sec. 32, T. 17 N., R. 9 E., 2 miles northeast of Nevada City. Elevation 3000′.
Bibliography: Cal. State Min. Bur. Repts. XI, page 300; XII, page 202; XIII, page 240. U. S. Geol. Survey Folio 29, Banner Hill Special.

**Yosemite.** (See Albert.)

**You Bet Mining Company.** Owner, Birdseye Creek Mining Company, Nevada City; Geo. Wight, Nevada City, president; C. P. Banning, secretary.

Location: You Bet Mining District, Secs. 6, 26, 28, 30, 31, 32, 36, 25, and 36, Twps. 15 and 16 N., R. 10 E., 7 miles northwest of Dutch Flat. Elevation 2600-3200′.

(See under Hydraulic Mining for activity planned in 1918.)

## GOLD—HYDRAULIC MINES.

At North Bloomfield and North Columbia hydraulic mining was developed and prosecuted on a very large scale from 1865 to 1880. The immense amount of gravel discharged into the Yuba River from the hydraulic mines operating from North Bloomfield to French Corral was the main cause of the agitation against hydraulic mining by the inhabitants of the Sacramento Valley, which culminated in the suspension of this method of mining in 1894.

The largest operations were carried on at North Bloomfield where from 1866 until the closure of the mine, February 1, 1894, the production had been $2,829,869. From 1894 until 1900 the property was worked at intervals and the total production to date is said to have been about $3,500,000 from 30,000,000 cubic yards of gravel. The channel of the Neocene river at this point was from 500′ to 600′ in width. The bank washed reached a maximum height of 500 feet, but most of the gold was obtained from the first 150′ above bedrock. This gravel averaged over 10¢ per cubic yard.

At North Bloomfield it was necessary in order to secure sufficient grade, to drive a tunnel from Humbug Creek 8875 feet in length. This tunnel was completed only a short time before the adverse decision in the suit of *Woodruff* vs. *The North Bloomfield Gravel Mining Company*, practically stopped hydraulic mining in the Sierra Nevada. Following the formation of the California Debris Commission by act of Congress in March, 1893, hydraulic mining was resumed on a small scale at North Bloomfield. The increasing severity of the restrictions passed by the commission finally resulted in a complete cessation of work in 1900.

From 1866 to 1900 about 30,000,000 cubic yards had been worked in the vicinity of North Bloomfield and the total production therefrom was approximately $3,500,000. It is estimated that 300,000,000 cubic yards, which will average over 10¢ per yard, still remain available for hydraulic mining in the North Bloomfield-Lake City area.

Near North Columbia the auriferous gravel deposits are extensive, owing to the junction of two of the large Tertiary streams near this point. The gold-bearing gravel at this point is from 400 to 500 feet in depth. It is estimated that 165,000,000 cubic yards still remain to be worked on the holdings of the Eureka Lake Company, which cover an area of 1445 acres and a distance of 2¼ miles along the channel. In the former operations only 25,000,000 cubic yards of the top gravel had been removed. It was planned to work the lower and richer gravels by means of a long bedrock tunnel, but on account of the suspension of hydraulic mining this was never done. The remaining gravel is said to run from 10¢ to 25¢ per cubic yard.

West of North Columbia the channel of the Tertiary Yuba River turned abruptly to the North. Beyond Badger Hill it has been eroded by the present Middle Yuba River, which has cut down into the granodiorite to a depth of 1000 feet below the bottom of the old Neocene stream.

From North San Juan to French Corral, the ancient river was preserved and the gravel deposits of this area have been extensively worked. The channel here reaches 1000 feet in width and the gravel averages 150 feet in depth. The top gravels here yielded from 10¢ to 15¢ per cubic yard, while the average value is said to have been 30¢. It is estimated that 52,500,000 cubic yards have been washed and that 12,500,000 cubic yards are still available. The production of the property to 1884 was $3,412,509.

From 1870 to 1882 over $4,000,000 was expended by the North Bloomfield Gravel Mining Company, its subsidiary companies, and the Milton and Union Mining Companies, in opening up the deposit, construction of reservoirs and ditches, and finally in driving bedrock tunnels.

The Bowman's Lake and other reservoirs, together with the 100 miles of distributing ditches, which were constructed at a cost of from $8000 to $10,000 per mile, are now owned by the River Mines-Eureka Lake companies.

It has been estimated that between $75,000,000 and $100,000,000 in gold lies buried in the gravels of the Neocene streams between North Bloomfield and French Corral.

This gold could probably be recovered without damage to the navigable streams or agricultural districts of the lower valleys, by using modern engineering methods in the construction of restraining dams in the lower reaches of the Yuba River at different points where natural dam sites exist. The first cost of such restraining dams would be large and it seems only reasonable that financial aid be given by state and federal governments for their construction. Such works would assist materially in solving many of the problems now being faced by the agricultural, irrigation, and navigation interests.

**Badger Hill.** (See River Mines Company.)

**Baltic.** (See under Lode.)

**Beckman.** (See Mayflower, Lode.)

**Birdseye Creek.** (See You Bet Mining Company.)

**I. X. L. Claim.** Owners, James Howlett, North Columbia; H. C. Dillon Estate, 2850 Leeward Avenue, Los Angeles.

Location: North Bloomfield Mining District, Secs. 3 and 10, T. 17 N., R. 9 E., 3 miles west of North Bloomfield.
Bibliography: U. S. Geol. Survey Folio 66; Colfax. U. S. Geol. Survey Prof. Paper 73, W. Lindgren, Tertiary Gravels of the Sierra Nevada.

The I. X. L. placer claim is an elongated area of about 100 acres situated a mile northwest of Lake City. The famous North Bloomfield tertiary channel is in part covered by this claim, but owing to the suspension of hydraulic mining no work has been done for many years. For description of the channel see River Mines Company.

**Liberty Hill Mines.** Owners, Wm. Maguire and Phœbe Maguire (one-half), and Anna F. Smith (one-half); lessees (with option to purchase) W. S. Bliss et al., 611 Insurance Exchange Building, San Francisco. The properties consist of the Liberty Hill Placer Mine (patented), Lots 41 and 66 in Sec. 23, T. 16 N., R. 10 E., 494.97 acres; the Maguire placer locations, containing about 160 acres in the same section; the Little York Placer Mine (patented), which includes Lots 37 and 38 in Sec. 5, T. 15 N., R. 10 E.; and Lots 39, 40 and 76 in Sec. 33, T. 16 N., R. 10 E., 430.9 acres; an undivided one-half interest in

Photo No. 10.    Liberty Hill Hydraulic Mine.    Photo by C. A. Logan.

the Consolidated Junction Placer claim in Secs. 26, 27, 34, 35, T. 16 N., R. 10 E., in Placer County, 75.41 acres; and an undivided one-half interest in the Liberty Hill and Polar Star Tailings Dam in Bear River, half a mile from Dutch Flat, being in all 1161.28 acres.

Location: The Liberty Hill property is about 3 miles by trail or 7 miles by road from Dutch Flat on the opposite side of Bear River, at an elevation of 3349 feet. The Little York Mine is about 4 miles southwest of Dutch Flat by road.

Lindgren[*] regarded the Little York deposit as pretty well worked out, but estimated that of a total of 18,000,000 cubic yards at Liberty Hill only 1/9 had been hydraulicked.

The channel is said to be 600 yards wide at Liberty Hill. The gravel is 60' to 85' deep and without overburden. It contains some very heavy boulders of gabbro and related rocks and is generally loose. The

[*]Lindgren, W., Tertiary Gravels of the Sierra Nevada of California: Prof. Paper 73. U. S. Geol. Surv., pp. 144–146.

bedrock in this section of the old stream, rises rapidly. The property is equipped with 45 miles of ditch, the longest being 9 miles long and delivering the water under 250 feet head. The ditches are largely intact, but will require considerable cleaning, as will also the reservoir. The main pipe line into the workings is apparently in good condition. The property has first rights to water out of Bear River, and the normal supply will permit six months mining.

Photo No. 11. Log crib, hydraulic fill dam, formerly used to restrain tailings from Liberty Hill and Polar Star hydraulic mines, being raised in November, 1918, preparatory to renewed activity at former mine. Located in Bear River one-half mile from Dutch Flat. Photo by C. A. Logan.

A force of about twenty men were at work in September, 1918, raising the dam on Bear River, preparatory to hydraulicking the Liberty

Photo No. 13 A.   Chinamen cleaning bedrock at the Omega
Hydraulic Mine.

Hill property the following winter.   Hydraulic mining with one giant
and about 1200 inches of water began January 15, 1919.

**Manzanita.**   (See under Drift.)

**Mayflower.**   (See under Lode.)

**North Bloomfield.** (See River Mines Company.)

**Odin.** (See Manzanita, Drift.)

**Omega Mine** (Prescott). Owner, Omega Placer Mining Company, 1213 Fischer Building, Chicago, Illinois.

>Location: Washington Mining District, Secs. 16 and 17, T. 17 N., R. 11 E., 3 miles southeast of Washington. Elevation 4000'.
>Bibliography: Cal. State Min. Bur. Rept. XIII, page 258. U. S. Geol. Survey Folio 66. U. S. Geol. Survey Prof. Paper 73, page 147.

The Omega hydraulic mine was worked for many years before and after the restriction of hydraulic mining. The accompanying photograph will give a good idea of the amount of material which has been

Photo No. 14. Debris restraining dam of the Omega Hydraulic Mine near Washington. Photo by C. A. Logan.

worked.  About 13,000,000 cubic yards have been removed which is said to have averaged 13½ cents per cubic yard, and it is estimated that 40,000,000 cubic yards remain which can be worked.  In late years, up to about 1915, most of the operations have been carried on by Chinamen under a leasing system.

The gravel deposit is a portion of an old Neocene river and the ground is 175 feet in depth.  The elevation of the bedrock is 4028' or 1000 feet above the bed of the South Yuba River.  There are two strata of gravel.  The bottom 150 feet is small gravel with a large amount of quartz, the greater part of which does not exceed 6 inches in size, although some large boulders of granite are included.  Above this there was from 6 to 10 feet of fine pipe clay, overlaid by another 20 feet of fine gravel, which extends in a southeasterly direction under a capping composed of tuffs and volcanic breccia of andesite.  The bedrock is Calaveras slate.  Twenty Chinamen were at work cleaning up bedrock when the mine was visited in 1914 but no information could be obtained from them regarding production, costs, etc.  A new restraining dam was being constructed below the 3000-foot drain tunnel.

Since 1915 the property has been worked by two partners during the water season.  The photo of the restraining dam shows that it is filled nearly to capacity and must be raised if more mining is done.  Raising the dam and mining in 1918 is planned.

**River Mines Company** (Eureka Lake and Yuba Canal Company). Owner, River Mines Company, care Geo. W. Starr, Grass Valley, California.

Location: Claims in French Corral, North San Juan, North Columbia and
   North Bloomfield Mining districts.  Elevation from 2000' to 4000'.
Bibliography: Cal. State Min. Bur. Repts.  U. S. Geol. Survey Folios 18
   (Smartsville) and 66 (Colfax).  U. S. Geol. Survey Prof. Paper 73, Tertiary
   Gravels of the Sierra Nevada, Lindgren.

A partial list of the claims now controlled by the River Mines and Eureka Lake Companies is as follows:

| Name | Location. mining district | Sections | Town-ship | Range |
|---|---|---|---|---|
| River Mines Company | French Corral | 12, 14, 24, 25, 35 | 17 N. | 7 E. |
| Bed Rock Tunnel Company | French Corral | 14 and 23 | 17 N. | 7 E. |
| Kate Hayes Placer Mine | French Corral | 25 and 26 | 17 N. | 7 E. |

| | Section | Township and range |
|---|---|---|
| Milton, placer mine, French Corral | 26 | 17N., 7E. |
| Badger Hill and Cherokee, placer mines, North Columbia | 1 | 17N., 8E. |
| | 36 | 18N., 8E. |
| American, placer mine, North San Juan | 6, 7 | 17N., 8E. |
| Sebastopol, Sweetland, Bloomfield Hydraulic Gravel, placer mines, North Bloomfield | 1, 2, 11, 12 | 17N., 9E. |
| North Bloomfield Gravel Mining Co., North Bloomfield | 1 | 17N., 9E. |
| | 35, 36 | 18N., 9E. |
| | 6 | 17N., 10E. |
| | 31 | 18N., 10E. |
| Consolidated, placer mine, North Columbia-North Bloomfield | 5, 7, 8 | 17N., 9E. |
| Central Gravel, placer mine, North Columbia-North Bloomfield | 5, 7, 8 | 17N., 9E. |
| Northern Gravel, placer mine, North Columbia-North Bloomfield | 5 | 17N., 9E. |
| Western Gravel, placer mine, North Columbia-North Bloomfield | 6, 7 | 17N., 9E. |
| Union Gravel, placer mine, North Columbia-North Bloomfield | 9, 10 | 17N., 9E. |
| Humbug Creek, placer mine, North Columbia-North Bloomfield | 12, 12, 14 | 17N., 9E. |
| Relief Hill, placer mine, North Columbia-North Bloomfield | 4, 9 | 17N., 10E. |
| Waukashaw, placer mine, North Columbia-North Bloomfield | 6 | 17N., 10E. |
| Cooke & Porter, placer mine, North Columbia-North Bloomfield | 6 | 17N., 10E. |
| Snow, placer mine, North Columbia-North Bloomfield | 13, 14 | 18N., 10E. |
| Hazard, placer mine, North Columbia-North Bloomfield | 16, 21 | 18N., 10E. |

The above claims control the major portion of the Great Neocene Yuba River from North Bloomfield to French Corral, a distance of approximately 15 miles. This ancient stream emerges from beneath an andesite lava cap a mile north of North Bloomfield and from this point the gravel deposit is continuous until Badger Hill is reached, eight miles to the west.

From North Bloomfield to North Columbia the channel is from 300 to 600 feet in width and the gravel deposit from 150 to 500 feet in depth. Portions of the deposit are covered by sand and light-colored clays. Above North Bloomfield the main channel branches; one branch continues eastward toward Relief Hill and the other northward toward the Middle Yuba River. These channels lie under a capping of lava and the easterly or Derbec channel has been worked by drifting for a distance of 7000 feet up stream.

**You Bet Mines.** Owner, You Bet Mining Company; lessee (with option to purchase), California Placer Mining Company, W. L. McGuire, secretary, tenth floor Crocker Building, San Francisco.

The You Bet Mining Company started hydraulic mining in the winter of 1913–1914, after building a debris dam and getting a permit from the Debris Commission. The mine was worked successfully till August, 1914, when work was stopped by a Court injunction, it being claimed that the water, which was used below the dam for domestic purposes, was being rendered unfit for use because of its turbidity. In August, 1915, many of the claims were leased to Chinese, who began drift mining near the old 'tie' workings near You Bet.

The present lessees took the property in the summer of 1918. The dam in Missouri Cañon is to be raised 15 feet. It is a gravel dam with concrete spillway. The property has 15 miles of ditches and a maximum water supply of 1200 inches. There is said to be about ¾-mile of unworked channel between the old Hayward shaft and the Nevada tunnel. There are three pipe lines into the workings under heads of 100, 200, and 300 feet respectively.

The work will be done mostly in cemented gravel, which is from 60 feet to 100 feet thick. Powder drifts and crosscuts will be driven, loaded with powder and shot to break up the cement. Probably three giants with 5″ and 6″ nozzles will be used in piping and a 5″ giant will be used to stack tailings. The property is equipped with 900 feet of 6-feet sluice split into two 5-feet sluices at the pit and with iron rails and block riffles. Twenty men were employed in September, 1918, preparing for mining. It is said that this mine has yielded 40¢ a yard in hydraulic mining and $7 to $9 a car in drift mining. It is estimated that 50,000,000 cubic yards of gravel have been worked in the past and that 100,000,000 cubic yards are available. The You Bet hydraulic mines are said to have produced about $3,000,000, in addition to yield since 1913, for which figures are not obtainable. It is known that operations in 1913–1914 paid well, however.

The gravel of the deep channel in the vicinity of You Bet is coarse and cemented in the deep trough, but in the upper portions is fine quartz gravel mixed with sand. In the deep channel the gravel is from 75 feet to 200 feet deep, and on the benches from 90 feet to 100 feet, and the width of pay gravel is 300 to 400 feet.

The following is a list of the claims held by this company:

| | Section | Township | Range | Area |
|---|---|---|---|---|
| Birdseye Canyon placer mine | 6 | 15 | 10 | |
| | 31 | 16 | 10 | 49.65 |
| Walloupa Canyon placer mine | 6 | 15 | 10 | |
| Arkansas placer mine | 26 | 16 | 10 | |
| | 25 | 16 | 9 | 35.60 |
| Greenhorn placer mine | 30 | 16 | 10 | |
| | 6 | 15 | 10 | |
| Washington placer mine | 31 | 16 | 10 | 37.94 |
| Red Dog placer mine | 30 | 16 | 10 | 48.03 |
| You Bet placer mine | 31 | 16 | 10 | 107.01 |
| Neece & West, Brown's Hill | 6 | 15 | 10 | |
| Walloupa | 31 | 16 | 10 | 87.59 |
| Starr | 25 | 16 | 9 | 24.84 |
| | 30 | 16 | 10 | |
| Mallory and other claims | | | | 5.37 |
| Brown Bros. placer mine | 6 | 15 | 10 | 28.65 |
| | 31 | 16 | 10 | |
| Poverty placer mine | 6 | 15 | 10 | 60.86 |
| | 31, 32 | 16 | 10 | |
| Benj. F. Myers claim | | | | |
| Placer mining claims in | 28 | 16 | 10 | 151.29 |
| Missouri Canyon placer mine | 36 | 16 | 9 | |
| | 30 | 16 | 10 | 19.51 |
| Newark Fluming and Mining Co. placer mine (Greenhorn Creek) | | 15 | 9 | |
| | | 16 | 10 | |
| Live Oak (½ interest in G. Atkins and J. F. Taylor placer mine) | 31 | 16 | 10 | 7.64 |
| Total | | | | 663.38 |

## GOLD—LODE MINES.

**Ajax Mine.** Owner, Miss J. Mitchell, Grass Valley.

Location: Grass Valley District, Secs. 2 and 11, T. 15 N., R. 8 E., 3 miles south of Grass Valley. Elevation 2200'.

This property consists of one patented claim (20 acres) adjoining the Allison Ranch mine on the south. The vein, which strikes north and dips 45° W., has an average width of 1 to 2' and can be traced for a distance of about 500' on the surface. It is the southern extension of one of the Allison Ranch veins and occurs in the granodiorite. The shallow drifts and tunnels driven from the banks of Wolf Creek are at present inaccessible.

**Alaska Mine.** Owner, Alaska Mining Company; M. S. White, Grass Valley.

Location: Grass Valley District, Sec. 1, T. 15 N., R. 8 E., 2½ miles southeast of Grass Valley. Elevation 2500'.

In January, 1916, the mine was being operated by the people who are working the Ben Franklin: J. L. Clayborn, representative. After six months work it was closed again and is idle, September, 1918.

Photo No. 14 A.   Surface plant of the Alcalde Mine (better known as the Kenosha or Seven Thirty).

**Alcalde Gold Mines Company** (also known as Bowery, Seven-Thirty, or Kenosha). Owner, Bailey-Drake Company, E. E. Drake, president, and L. A. Bailey, secretary. Office, 149 New Montgomery street, San Francisco.

Location: Grass Valley Mining District, Secs. 5 and 6, T. 15 N., R. 8 E., 4 miles southwest of Grass Valley.

There are twelve claims, known as Eclipse, Hidden Treasure, Accidental, Oxford, Diamond, Philadelphia, Stonewall Jackson, Jonson, Gould and Curry, South Curry, Indiana and South Standard, covering 200 acres in area and 4500 feet along the course of the lode. Applications for patents were made a few years ago, but these were delayed by a contest filed by the Central Pacific Railroad.

The principal vein so far exploited occurs in a fissure in amphibolite probably derived from diabase, gabbro and diorite. The width of 'formation' is from 18" to 2', but the width of quartz averages about 6". The strike is N. 17° W.; and the dip varies from 45° to 75° W. The ore is free milling and the gold generally coarse, with sulphides running from 2% to 3%. The owners emphasize the similarity of the vein system to those in the same district farther east, and claim that the same increased richness has been found in the neighborhood of cross fissures, or 'iron-crossings.'

The development work has been carried on along the vein on the Hidden Treasure, Eclipse and Gould and Curry claims. The first work on the claims was done in 1867 by three Frenchmen who obtained coarse gold in sluicing the bed of French Ravine as far as the mouth of a small ravine, where the Kenosha vein was found. The first day's work on the outcrop was well repaid. In October, 1867, the Bowery shaft had been sunk 90 feet and a drift had been driven 70 feet on a vein yielding crystallized gold. This shaft was continued to water level (130 feet on the dip of the vein), and the reported yield was high, but the amount of lateral development is not known. About 250 feet north of the Bowery shaft, the McCook shaft was sunk 180 feet on the dip of the vein and a level was driven north from a depth of about 120 feet. Ore was stoped for a short distance above the drift apparently as far as the Taylor shaft, about 250 feet north of the McCook. Work continued through the year 1872, and the files of the Grass Valley Union indicate the production during the five-year period of a number of small but very rich lots of ore. The property was worked by individuals, or small partnerships, and no systematic records were kept. The Taylor shaft was sunk about 180 feet during this period, opening up a vein 42 inches to 52 inches wide. A drift was driven north 100 feet and the longitudinal section map indicates considerable stoping near the shaft. All the operations during the old days were limited to the zone above water level.

In 1875 the claims were relocated by Richards and Desmond, passed to Geo. W. Root, and later to the present owners. The old shafts are now inaccessible. In 1900 one Riley got permission to sink and put down the Riley shaft on the southern part of the Hidden Treasure claim and is said to have taken out $24,000 in working to a depth of 180 feet.

The Taylor shaft has been used as a working shaft in later work. It is a double compartment shaft and in June, 1918, had been sunk 450 feet on the dip of the vein. A second level was run north from a depth of about 170 feet for 200 feet on the vein. The third level, 100 feet below the second, was driven north 400 feet and south 200 feet. A raise was put through to the second level 100 feet north of the shaft, and a little stoping done. The second raise, 350 feet north of the shaft, was put through to the old Riley shaft. This raise is said to have proven the downward extension of the Riley pay shoot. The fourth level 100 feet below the third, was driven 325 feet north and 180 feet south on the vein. A report by Geo. W. Root indicates that the Riley shoot was encountered in the north drift and that several thousand dollars were produced from an area 10 feet long by 8 feet on the dip of vein. Since spring the workings have been allowed to fill with water, which stood 108 feet below the collar of the shaft in September, 1918.

The buildings include mill building, hoist and compressor room, blacksmith shop, shaft house, change room, office, etc. The equipment comprises a five-stamp mill, six drill Giant air compressor, Cornish pump, steam hoist and steam pump system, several pumps and machine drills. There are three electric motors to drive the compressor, the Cornish pump and the stamp mill. Geo. W. Root, the former owner, estimates the total production of the property at about one-half million dollars.

**Allison Ranch.** Owner, Grass Valley Consolidated Gold Mines Company.

Location: Grass Valley District, Sec. 2, T. 15 N., R. 8 E., 2¾ miles south of Grass Valley. Elevation 2150'.
Bibliography: Cal. State Min. Bur. Repts. XI, page 268; XIII, pages 2-4. U. S. Geol. Survey, Nevada City Special Folio.

The Allison Ranch mine was discovered during placer mining operations in Wolf Creek in the early '50's. The property embraces an area of about 64 acres of irregular outline, covering about 2400' of the Allison Ranch vein. The mine was located in December, 1854, and from 1854 until 1866 was one of the most famous producers of the Grass Valley district, yielding $2,300,000 from 46,000 tons of ore and paying during this period $1,200,000 in dividends. The vein at this time had been worked by an inclined shaft to a depth of 475' and four levels had been driven about 700' along the vein. The mine was closed in October, 1866, but was again reopened in April, 1869. From April, 1869, to December, 1871, the shaft was deepened to 600', a drift was run for a distance of 500' south of the shaft, and the production for the 2½ years is said to have been from $200,000 to $250,000.

The mine was idle from December, 1871, until 1896, when the property was purchased by Mackey & Flood of San Francisco. It was reopened and after installing an extensive plant the inclined shaft was continued to a depth of 1650', but instead of following the vein, the shaft was continued in the granodiorite at a uniform inclination. At a depth of 800', 1000', 1200', 1400' and 1600', crosscuts were driven east from the shaft intersecting the veins on each level. Drifts were then driven on the vein, an average distance of 400' north and 1000' south. Over two miles of drifts were driven on the Allison Ranch, Cariboo and branch veins, from 1896 to 1903. In the fall of 1902 a 20-stamp mill was completed and the yield for four months is given as follows:

| | Bullion | Concen- trates | Total |
|---|---|---|---|
| December, 1902, 10 stamps_____ | $4,750 | $1,410 | $6,190 |
| June, 1903, 10 stamps_____ | 5,400 | 1,530 | 6,980 |
| February, 1903, 20 stamps 25 days_____ | 10,800 | 3,060 | 13,860 |
| March, 1903, 20 stamps 20 days_____ | 10,650 | 2,970 | 13,620 |
| Totals _____ | $31,600 | $9,000 | $40,600 |

The ore during this period is said to have averaged $10 per ton milled and the concentrates $80 per ton.

The operating costs per month were approximately as follows:

Power _____ $5,000
Labor underground _____ 4,500
Labor on surface_____ 1,900
Labor at mill_____ 550
Supplies _____ 1,000

Total _____ $12,950

Owing to the fact that the shaft was sunk near Wolf Creek, 1000 gallons of water per minute had to be handled which accounts in part for the high power cost.

Mr. John W. Mackey, who was the leading spirit of the enterprise, died in London during the summer of 1902 and following his death a complete examination of the property was made by Ross E. Browne. Shortly after this report was rendered and the test mill-run completed the mine was closed, the entire equipment dismantled and sold, and the property has remained idle ever since. It is currently reported that a considerable amount of $10 ore was blocked out in the mine when it was closed, but it seems improbable that the mine, completely equipped and with a new 20-stamp mill in operation, would have been abandoned if such had been the case.

The Allison Ranch vein belongs to the Omaha-Wisconsin-Hartery north-south system of parallel veins, occurring in the granodiorite area.

Photo No. 15. The Allison Ranch Mine in 1914, before the erection of the new plant.

It varies in strike from N. 0° to N. 15° E. and dips 45° W. The outcrop of the vein is inconspicuous, but it has an average width underground of about 18' down to the 600' level, where it is reported to have 'broken up.' The vein below this level as developed by Mackey averages 2' with a maximum width of five. The Allison Ranch vein as it was followed south is said to have 'split' and the vein which was followed in the old workings in a southeasterly direction was called the Cariboo Ledge. This was a small vein varying from 4" to 10" in width, but yielding rich ore running from $200 to $300 per ton.

The granodiorite which forms both walls of these veins shows considerable sheeting and there are a number of parallel veins lying in the hanging wall of the Allison ranch ledge. The only attempt that was made to explore these veins was by a diamond drill hole which was driven 100' west from the 800' level, but was discontinued after a heavy flow of water was encountered. The quartz in these veins carries free gold and from 2 to 5% of sulphides consisting of pyrite, galena and a small amount of chalcopyrite.

[When the mine was visited in September, 1918, the workings had been unwatered to a depth of about 900 feet, and the superintendent reported that new drifts totaling about 1000 feet had been driven, as well as a crosscut several hundred feet long. Some work has been done on each level from the third to the eighth since the mine was reopened. The inclined shaft is 1675 feet deep and there are 11 levels in the mine. The problem of handling the water is rather difficult, but not necessarily insurmountable. New work is being carried on in the direction of the Cariboo vein on the 300-ft. and 400-foot levels. New ore was being milled. The vein is said to vary from 8 inches to 15 inches in width, and the cost of mining and milling, not counting development, is placed at about $5 a ton.

Ross Browne's report showed the presence of ore ranging from $1 to $24 a ton and the average value of all assays, according to his figures, was such that successful operation ought to be possible under careful management and favorable conditions. Browne believed that the vein had been displaced by a number of step faults at different depths, so that while it might be lost in the course of sinking on it, it could be picked up by crosscutting from a hanging-wall shaft.

The surface plant at the mine includes a 102-ft. wooden headframe, a hoist capable of handling 2 tons from an incline depth of 2500 feet, a new Sullivan compressor of 1500 cubic feet free air capacity and an Ingersoll-Rand compressor of 500 cubic feet capacity, and shop.

The mill is equipped with 20 stamps weighing 1450 lbs. each, mounted on 10,000-lb. mortars. Sulphides are saved on 5 Frue vanners. There is a new cyanide plant of 100 tons capacity which has been run only 2 or 3 weeks, and was not in use when visited.  Ore has to be hoisted from the bin in the headframe to the mill.  Here it goes onto a shaking table where it is washed and waste is hand picked before the ore is broken in the Gates crusher.  Concentrate is said to run about $40 a ton.  A total force of 45 men are employed.—*C. A. Logan.*]

**Alpha Mine.**  (See Golden Gate Group.)  Owner, Wm. G orge Estate; A. G. George, Grass Valley.  Under option to Golden Gate Consolidated Mine Company, Chas. C. Haub, 45 Powell Street, San Francisco.

Location: Grass Valley District, Secs. 25 and 26, T. 16 N., R. 8 E.

**Alpine Mine.**  Agent for owner, J. T. Morgan, Nevada City.

Location: Nevada City-Banner District. Sec. 17, T. 16 N., R. 9 E., 2 miles east of Nevada City.  Elevation 2700'.

The property consists of two patented claims, the Alpine and Little Dove, containing about 40 acres and covering approximately 1400' of the eastern extension of the strong quartz-filled fissure known as the St. Louis vein.  This vein, which strikes northeast and dips 75° N., is from 1' to 4' in width and carries free gold, pyrite, and some galena and sphalerite.  The development consists of a tunnel driven 400' on the vein, attaining a depth of 100' below the outcrop, and a shaft 125' in depth sunk on the vein on the Alpine claim.  The Wide West vein has a north strike and westerly dip; it intersects the St. Louis vein on the Little Dove claim.  Very little exploratory work has been done on this ledge, and no ore bodies of value have been developed.

**Ancho Mine.**  Owner, George Mainhart, Grass Valley.

Location: Graniteville District, Secs. 16 and 21, T. 18 N., R. 11 E., 2 miles southeast of Graniteville.  Elevation 2000'.

The Ancho mine was first located about 1878 by Gashwieler, and was purchased by the present owners in 1908.  There are two patented claims: the Ancho (2300' x 200') and the Edison (700' x 600'); and four locations, the West Virginia, Wheeling, Ohio and Nevada, a total area of 100 acres, covering 3500' along the Ancho lode.  There are four quartz-filled fissures on the property, the West Virginia with an average width of 4', the Wheeling with 50' of quartz and slate, and the Ancho and Ohio, each about 4' in width.

The general strike of the veins which occur in Calaveras slates near their eastern contact with the granite, is a few degrees west of north, and the dip is from 60° to 70° E.  The quartz carries free gold with as high as 3% of pyrite and galena.  Most of the development work so far undertaken has been on the Ancho ledge, which was intersected at a depth of 200' below the outcrop by a 250' crosscut adit.  A drift was then driven on the vein 700' towards the south, but no work has been done on the vein north of the crosscut. The adit tunnel which has been continued east of the Ancho vein a distance of 200' is expected to cut the West Virginia vein within 300' from the present face of the crosscut.

Another crosscut adit, which is being driven 600' below the upper workings to intersect the Wheeling vein at a distance of 600', is now in 200' (1914). The Ancho vein will also be opened at a depth of 250' below the upper workings by a 600' crosscut tunnel, which is now in a distance of 300'. The Ancho vein has been stoped for a length of 600' and to a maximum height of 100'. The ore (as is characteristic of this district) is low grade.

The mine is reported to have produced about $50,000 to 1914.

Equipment consists of a 10-stamp mill. Rix 12" compressor, blacksmith shop, boarding house, and office. Water is obtained for power from the North Bloomfield ditch under 400' head at a cost of 10¢ per inch. Five men were employed during the summer of 1914.

The property was shut down in August, 1918, because of high costs and adverse working conditions.

**Arkansas Traveler.** (See under Copper.)

**Arctic Mine.** Owner, Arctic Mining and Power Company, 538 Consolidated Realty Building. Los Angeles; J. T. Price, president, Los Angeles; J. P. Flint, manager, 2019 W. Washington Street, Los Angeles.

Location: Washington District, Sec. 2, T. 17 N., and Sec. 35, T. 18 N., R. 11 E., 7 miles northeast of Washington. Elevation 3600'–5500'. Bibliography: Cal. State Min. Bur. Rept. XIII, page 235.

The claims comprising this group are the Washington Chief Nos. 1 and 2, patented, and locations known as the Cañon Mascot, Arctic, North Arctic, South Arctic, West Arctic, Golden Rule, North Golden Rule. Good Luck, Sirus and Sirus Extension. The total area held amounts to about 228 acres and covers about 9000', along the strike of the Arctic vein. There are on this property eight parallel ledges belonging to the Eagle Bird-Ethel vein system, but of these the Arctic, and the Golden Rule (lying 600' east of the Arctic) are the most important. These fissure veins occur in granite, and the Arctic varies in width from 1' to 30' with an average width of 5', while the Golden Rule averages about 15' in width. The strike of both veins is a few degrees west of north, but the Arctic vein dips from 55° to 65° E., while the Golden Rule dips 60° W., and it is expected that the veins will intersect at about the elevation of the lower tunnel level. The character of the ore also differs in the two veins; the Arctic ore contains only a small percentage of sulphide associated with the free gold, while the Golden Rule ore carries as high as $7\frac{1}{2}$ per cent of pyrite, galena and zinc blende.

The development work on the Arctic lode consists of a shaft 60' in depth and two tunnels which have been driven on the vein 260' and 30' at an elevation 700' above the lower tunnel. Work is now being confined to the driving of the main crosscut tunnel from Cañon Creek, elevation 3750' which has been driven 140' and will at a distance of 965' intersect the Arctic-Golden Rule ledges, 600' below the upper workings. The only work on the Golden Rule vein is an open cut.

An up-to-date and complete hydro-electric power plant was completed in 1914. The plant includes a Pelton Francis 25" turbine direct connected to a 600 k. w. 6600 V., 400 r. p. m., 3 phase, 60 cycle generator with transformers, all housed in a reinforced concrete building. The company controls a two thousand inch water right on Cañon Creek; the water is at present to be used under a head of 165 feet, but later by means of a 2-mile ditch and flume, this amount of water will be obtained under a 700' head. A new office building and superintendent's house have just been completed. It was the intention of the company to install an air compressor in 1915 to be later followed by a complete reduction equipment which was to include the Reid Electric Smelting process for the treatment of concentrates. Since the above was written, the compressor has been installed and a few hundred feet of drifts are said to have been driven, without encountering the vein. Idle in September, 1918.

### Atlantic Claim.

Location: 1½ miles southeast of Eureka, in T. 18 N., R. 11 E., M. D. M.
Bibliography: Cal. State Min. Bur. Rept. XIII, page 235.

### Badger Hill Claim. Owner, River Mines Company; Geo. W. Starr, manager, Grass Valley.

Location: Sec. 1, T. 17 N., R. 8 E., and Sec. 36, T. 18 N., R. 8 E., at Cherokee, Nevada County.
Bibliography: Cal. State Min. Bur. Repts. XII, page 185; XIII, page 235.

See River Mines Company (Hydraulic).

### Bagley Mine. Bonded to Clyde S. Carr of Nevada City and Howard Dennis of Grass Valley.

Location: Nevada City Mining District.
Bibliography: U. S. Geol. Survey, W. Lindgren, Prof. Paper 73, pages 125–132. U. S. Geol. Survey 17th Annual Rept. pt. I, pages 1–262, 1896. U. S. Geol. Survey Folios 18 and 29, Smartsville and Nevada City.

This property, taken under bond by Carr and Dennis in September, 1915, has yielded some rich ore, but has only been worked along limited lines. The shaft has been retimbered down to a former tunnel level and retimbering of the old workings commenced. It is planned to sink the shaft to a depth of 500' and prosecute lateral work. Early in October, 1915, a 5' vein assaying $30 to $40 was encountered while following a small stringer. Development was proceeding to determine the dimensions of the ore-body. This property was idle in September, 1918, and the last operator was not in town at time of visit.

**Baltic Mine.** Owner, Mary E. Shaser, 36 Weeks avenue, Santa Cruz; A. Maltman and J. M. Bean.

Location: Washington District, Secs. 26, 27, 34, and 35, T. 18 N., R. 11 E., 7 miles northeast of Washington.
Bibliography: Cal. State Min. Bur. Repts. VIII, page 451; XII, page 185; XIII, page 235. U. S. Geol. Survey, Colfax Folio.

The property includes the Crown Point, Crown Point Ext., and Shirley patented claims controlling 4500' along the strike of the vein, which dips 45° E. This vein belongs to the Eagle Bird-Yuba-Ethel-Birchville vein system occurring as a fissure vein in the granite. The mine has been worked at intervals and at one time a 20-stamp mill was in operation. Idle in 1914. Idle in September, 1918.

**Baltic Mine.** Owner, Baltic Gravel Mining Company; Dr. I. W. Hays, president, Grass Valley.

Location: Rough and Ready District, Secs. 20, 29 and 30, T. 16 N., R. 8 E., 3 miles west of Grass Valley. Elevation 2000'.

Located in early days and worked as a hydraulic mine, this property, consisting of 147 acres of patented placer ground and a 20-acre location, was purchased by the present owner 40 years ago. In the past few years some work has been done on two parallel quartz veins about 600' apart. The strike of these veins is northwest and the dip is from 60° to 70° W. The veins occur near the contact of the amphibolite and gabbro and the ore contains free gold and pyrite. The property has recently been under bond to the operators of the Gold Mound mine, which lies about one-half mile east of the Baltic. Idle in September, 1918.

**Beckman Mine.**

Bibliography: Cal. State Min. Bur. Repts. VIII, page 453; XI, page 295; XII, page 192; XIII, page 252.

See Mayflower.

**Belle Fontaine Mine.** Owners, Belle Fontaine Mining Company; R. M. Martin, Nevada City, and W. J. Smith, Nevada City.

Location: Nevada City-Banner District, Sec. 9, T. 16 N., R. 9 E., 2¼ miles east of Nevada City. Elevation 2800'.
Bibliography: Cal. State Min. Bur. Repts. XII, page 185; XIII, page 235. U. S. Geol. Survey, Nevada City Banner Special Folio.

The Belle Fontaine patented claim is 1200' x 350' and was located in 1857. The vein is well defined, varying in width from 4 inches to 2 feet with an average width of 1 foot. It strikes E. and dips 20° to 30° N., occurring in the granodiorite about 1000' west of its contact with the Calaveras slates. The mine has been worked at intervals by means of tunnels driven from Deer Creek and a number of small ore shoots were developed. The Federal Loan and Lecompton are neighboring properties.

**Belle Union Mine.** Owners, E. A. Dunkley et al., Grass Valley.

Location: Grass Valley District, Sec. 11, T. 15 N., R. 8 E., 3 miles south of Grass Valley. Elevation 2000'.

The Belle Union was first located by James Butts about 1867, but was later abandoned and patented as railroad land. The present owners acquired title by purchase. The property, comprising an area of 80 acres, was last worked in 1910. The quartz vein, which averages about 2' in width, occurs in the granodiorite about one-half mile southeast of the Allison Ranch Mine. Its strike is N. and the dip 35° E. A tunnel was driven on the vein a distance of 500', but the depth attained thereby was only 50 feet. The ore was stoped from this level to the surface. A crosscut tunnel, said to lack about 50' of striking the vein, has been driven which will give 500' of backs. There is no equipment on the property and the workings are inaccessible at the present time.

**Ben Franklin Mine.** Owner, Ben Franklin Consolidated Gold Mining Company; M. S. White, Grass Valley.

Location: Grass Valley District, Sec. 1, T. 15 N., R. 8 E., 2¼ miles southeast of Grass Valley. Elevation 2600'.
Bibliography: U. S. Geol. Survey, Nevada City Special Folio.

The claims, comprising this property, include the St. Stephen, Cleveland, Ben Franklin, Wilcox, Avondale, Alaska, and others. There are two veins, the Alaska and Ben Franklin. The former, which is about 1' in width and dips 45° to the east, lies in the granodiorite about 700' west of the Ben Franklin vein, which latter strikes N., following the granodiorite-diabase contact and dips 35° W. Both of the veins have been worked at intervals, the Ben Franklin by tunnels and shallow shafts, and the Alaska by an inclined shaft to a depth of about 500', and they have in the past produced considerable rich ore carrying free gold, galena, and pyrite. The Ben Franklin vein varies in width from 1' to 4'.

A company was organized by Toronto men early in 1915 to reopen the property. A pump was started to handle the water which was 200' deep. Hoisting and building equipment are complete. The total production is said to have been $750,000.

In October, 1915, New York capitalists represented by J. D. Clayborn, took a bond on the property and commenced unwatering the Ben Franklin shaft, and in January, 1916, it was reported that the new 5-stamp mill would soon be in operation. The property was abandoned, however, within two months after unwatering the shaft. It is said that a short drift was run and a little ore stoped.

**Berriman** (Golden Center, Dromedary, Rock Roche Consolidated). (See Golden Center.)

**Betsy** (Orleans). (See Sultana.)

**Big Bend.** (See under Copper.)

**Big Blue.** (See Murchie.)

**Birchville Mine** (Wisconsin, Iowa, Independence). Owner, Birchville Mining Company, 607 First National Bank Building, San Francisco; John A. Bunting, manager; H. W. Sweet.

Location: Graniteville District, Secs. 10, 15, and 23, T. 18 N., R. 11 E., 1 mile southeast of Graniteville. Elevation 5000'.

This mine was worked by the present owners for eight years up to 1914. The property consists of the following patented claims: South Commercial, Commercial Birchville, Independence, Iowa, North Commercial, Belmont No. 1, Belmont No. 2, Union, Wisconsin and South Wisconsin; locations known as the Bald Eagle, Buckeye (Klondyke) and Empire claims, together with two placer locations, the total comprising an area of 320 acres and a length along the strike of the lode of 7000 feet. There are two principal veins on the property, the Birchville, and the Commercial which lies 150' east of the Birchville at one point, but rapidly diverges owing to its different strike. The Birchville vein strikes N. 45° E. and dips 45° SE., while the Commercial has a strike of N. 5° E. and an easterly dip of from 35° to 65°.

Both walls of the veins are granite and the ore bodies occur in lenses, varying in width from 1 to 6 feet. The ore is free milling, and contains less than one per cent of pyrite and galena. Development work consists of a shaft which has been sunk to a depth of 400' on the Birchville ledge; the 200' level was driven 414' north and 300' south from the shaft; the 300' level was driven 450' north and 300' south and the 400' level was extended a distance of 450' north and 10' south. From the 200' level on the Birchville vein a crosscut 1600' in length was driven to the Commercial lode and drifts were run 250' north and 250' south of the crosscut; three raises were put up on the Birchville vein from these drifts a distance of 106', 120' and 130'. Ore has been stoped from an ore shoot 175' north of the shaft from the 400' to the 200' level.

This shoot is said to have been 150' in length, but the tonnage and value of the ore extracted could not be ascertained. The property is equipped with a 12 horsepower air hoist, and a 3-drill compressor driven by water power which is only available at present for about four months of the year. Surveys, however, have recently been completed for a ditch from Poorman Creek, which will give a better supply of water. The reduction equipment consists of two Straub mills which, crushing to 40 mesh, have a capacity of 8 tons each. Eight men were employed at the mine during the summer of 1914. The Rocky Glen mine lies about one mile west, and the Gaston mine 2½ miles southwest of the Birchville property. The property was reported to be idle in September, 1918.

**Bitner.** (See under Copper.)

**Black Bear Mine** (Forlorn Hope). Owner, Black Bear Mining Company, care of H. C. Black, Rough and Ready.

Location: Rough and Ready District, Secs. 13 and 14, T. 16 N., R. 7 E., 5 miles northwest of Grass Valley. Elevation 1600'.

The holdings comprise about 400 acres.

The Black Bear vein averages 3' in width, has a north strike and dips 55° to the west. It has been developed and worked by means of an adit tunnel driven into the south bank of Deer Creek. The mine has been worked at intervals in the past few years, and mill and tramways are installed. At the time the property was visited in 1914, however, it was idle and there was no one at the mine.

The new stamp mill was running on ore from the Forlorn Hope in February, 1914, and in Deecmber, 1914, the shaft was being unwatered preparatory to extensive development. In May, 1915, the Forlorn Hope vein was intersected at a depth of 175' disclosing an ore-body reported to be 3' wide, and by December a compressor, machine drills and air pump of 300 gal. capacity were installed, and the working force increased. W. C. Gans, superintendent.

[The mill was idle when visited in September, 1918, but two or three men had evidently been mining recently on the surface of the hillside just above the shaft. The mill contains a rock breaker, 10 light stamps on concrete foundations, discharging onto 4-ft. by 15-ft. plates, and two Johnson concentrators. There is also a Giant Air Compressor, and an amalgamating barrel. Water power is used and there are three Pelton water wheels, a small one for the concentrators, one about 5 ft. in diameter for the stamps and a larger one for the compressor.

Such ore as was in sight appeared to be from the broken, rusty stained upper part of vein and evidently came from superficial workings. The lower tunnel, near and on a level with the blacksmith shop, was caved at the portal; the dump here is all gabbro-diorite. The shaft is about ¼-mile from the mill and is sunk at an incline of about 45 degrees, having one compartment and manway. Just above the shaft on the hillside, another tunnel has been driven and a raise has evidently been put through the short intervening distance to the surface. The surface workings just above this showed some small quartz stringers.

At the shaft there is a Vulcan hoist operated by water power, a one-ton skip, truck and small ore bin. Ore is hand trammed to the mill.

It is said that short runs were made in March and June, 1918, and that small cleanups were made.—*C. A. Logan.*]

**Black Hawk Mine** (Aberdeen Tunnel Company). Owner, Theo. C. Dorsey, Grass Valley. Under option to Union Hill Mines Company, Crocker Building, San Francisco.

Location: Grass Valley District, Sec. 25, T. 16 N., R. 8 E., 2 miles east of Grass Valley. Elevation 2800'.

The Black Hawk mine was operated in early days by Smith and Canfield, and was purchased by the present owner in 1894. There are two claims, the Black Hawk (1500' x 400') which was patented by the Gold Point Company and then deeded to the Black Hawk, and the Rip Van Winkle fraction, a total area of about 18 acres, covering a length of 1500' along the lode. The Black Hawk veins which occur in the area of amphibolitic schists lying south and east of the Idaho-Maryland serpentine, belong to the Idaho-Maryland-Brunswick vein system. The vein upon which most of the work has been done is supposed to be the western extension of the Brunswick lode.

The vein filling consists of quartz carrying free gold and pyrite. Some fine 'specimens' have been taken from the oxidized zone above the Pike tunnel. The strike of the vein is N. 70° W., the dip 35° to 40° S., and the average width about 2 feet. Two tunnels have been driven on this property; the Pike tunnel which is a crosscut adit 620' in length cutting the vein 130' below the outcrop and the Aberdeen tunnel 1265' in length giving 450' of backs on the vein. The Pike tunnel was continued into the footwall a distance of 150' through a kaolinized zone, but no other veins were encountered. A drift was run 240' west of the Pike tunnel, in which a number of small irregular ore shoots were encountered, and a winze was sunk 35' west of the tunnel to a depth of 835 feet. No drifting was done to the east. The only development work done on the Aberdeen tunnel level was a drift which was run 40' west on the vein; at the present time this tunnel is inaccessible. While several tons of ore, which have been crushed as samples, have varied in value from $12 to $25 per ton, most of the stoping has consisted in following rich stringers and seams containing specimen ore. A 5-stamp mill is now being installed on the property. (No ore has been milled since the above was written, and the property was idle in September, 1918.)

**Black Prince.** (See Mountaineer.)

**Bluebell Group.** Owner, J. H. Von Schroeder, San Rafael.

Location: Washington District, Secs. 12 and 13, T. 17 N., R. 11 E., 5 miles east of Washington.

There are two claims, the Bluebell and Sunset. The property has been idle for a number of years.

**Blue Jay Mine.**  Owners, Geo. Bonney and M. H. Baugh, Nevada City.

> Location: Washington District, Sec. 2, T. 17 N., R. 11 E., 5 miles east of Washington. Elevation 4250'.
> Bibliography: Cal. State Min. Bur. Rept. XIII, page 236.

The Blue Jay mine is located on the steep divide which separates Cañon Creek from the South Fork of Yuba River. From the South Yuba, 3000' above sea level, there is a rapid ascent until the summit of the divide is reached at an elevation of 5000', less than a mile north of the river. The Blue Jay vein outcrops on the south side of the ridge at an elevation of 4250 feet. The vein, which strikes N. 20° W. and dips 65° E., has an average width of 4', and has been developed by means of a tunnel and winze, reaching a depth of about 300' below the outcrop. A 5-stamp mill was operated on this property at one time, but at present the mine is idle.

**Boss.**  (See under Copper.)

**Boston Ravine.**  (See North Star Mines Company.)

**Bowery.**  (See Alcalde Gold Mines Company.)

**Bowery Lodge.**  (See North Star Mines Company.)

**Brunswick Mine.**  Owner, Brunswick Consolidated Gold Mining Company, 519 California street, San Francisco.

> Location: Grass Valley Mining District, Secs. 25 and 36, T. 16 N., R. 8 E., 2 miles east of Grass Valley. Elevation 2600'.
> Bibliography: Cal. State Min. Bur. Repts. VIII, page 431; X, page 381; XI, page 274; XIII, page 237.

The Brunswick mine in the last few years has developed into one of the dividend-paying mines of the Grass Valley district. The property consists of an irregular-shaped claim containing 27 acres and covering 2900' along the strike of the lode; together with 320 acres of patented agricultural land, known as the Matteson ranch. The vein producing in 1914 belongs to the famous Idaho-Maryland vein system, but occurs in the amphibolite schist instead of along the serpentine contact as does the Idaho-Maryland vein.

The mine was located in early days, but in 1888 had only been worked to a depth of 300 feet. By 1896 the three-compartment shaft had reached a depth of 700', and later it was continued to the 1250' level and extensive lateral development work was undertaken.. The 1250' level was driven southeast on the vein for a distance of 2000', and has recently been connected with the new vertical shaft, at a depth of 875' below the collar. This new shaft cuts the vein at a depth of 975', and the total depth is now 1347 feet. On the 1250' level a new ore shoot was opened 1200 ft. east of the old inclined shaft; a raise was put up in ore a distance of 400', and drifts were run east and west therefrom. This pay shoot which was 500' in length has been stoped to a point 450'

Photo No. 16. On left, the hoist and mill of the Union Hill Mine. In center, near top of hill, is the old plant of the Brunswick Mine, and on the right is the new Brunswick plant.

above the 1250' level, at which place an intersection of two veins occurred causing an enrichment. In 1914 the ore from this shoot was being lowered to the 1250' level, trammed to the shaft, hoisted to the surface, trammed to ore-bins, reloaded into cars, trammed to the old mill and finally shoveled into the rock breakers. At the new shaft, which is located within a few feet of the Nevada County Narrow Gauge Railroad, a new hoist and Ingersoll-Rand compressor, both electrically operated, have recently been installed, together with machine and carpenter shops and other necessary buildings. In 1915, a steel head-frame and a complete 20-stamp mill and cyanide plant were installed, and the method of handling and treating the ore have thereby been greatly improved and the operating costs materially reduced.

The strike of the Brunswick vein is N. 50° W., and the dip varies from 40°, in the upper levels, to 70° SW. below the 700' level. As a rule the walls of the vein, which are from a few inches to five feet apart, are well defined. The filling between the walls is in places solid quartz, in other parts it is composed of altered schist and stringers of quartz. This material carries free gold, pyrite, galena, and some chalcopyrite. Ore of this character from the east pay shoot is said to have averaged $20 per ton.

[Since the above was written, operations were carried on continuously throughout 1916 and 1917. The general manager's report for 1916 showed that 2185 feet of underground work was done that year; and total production for the year was $196,521. The new 20-stamp mill went into commission in October, 1915. In January, 1917, there were about 30,000 tons of ore blocked out. In 1917, according to the general manager's report only 1896 feet of underground development was done, the temporary result of which was a lower total cost per ton. Just below the floor of the drift on the 900 ft. level, the vein split and the ore below the split was much lower grade than above. This zone of impoverishment had not been bottomed at 1347 feet, so its ultimate depth is uncertain. Both grade and quantity of ore are said to have decreased between the 1100 and 1300 ft. levels. About 10,000 tons of ore were produced from the levels above 900 feet, and no new ore was developed from these levels, which were considered exhausted. Very little ore was developed on the 1200 and 1300 ft. levels. A total of 30,805 tons crushed in 1917 gave an average yield of $5.92 a ton. At the same time there were advances in cost of casualty insurance, of all supplies except timber, and in wages. Overhead costs remained the same, with smaller tonnage, and provision had to be made for impounding tailings. Two-thirds of the tonnage was crushed in the new mill, extraction there being 90.8%. Ore from the upper levels carried

1.62% of concentrate worth $34 a ton; in ore from the lower levels the concentrate formed 2.01% and was worth $58 a ton.

Work in 1918 began with a very small ore reserve of low grade, and costs kept soaring. Shaft sinking had to be stopped at 1347 feet because the hoist and head-frame had reached their capacity, according to Mr. Turner. The mine was closed for the duration of the war in June, 1918. The company owns considerable unexplored ground under the Matteson Ranch, which is said to contain the extension of the Brunswick vein. Therefore, lateral exploration and sinking both offer promising possibilities which can best be exploited when conditions have returned to normal. The case of the Brunswick is typical of a number of properties, and the causes operating there can be seen at work today in the very best mines of the district. High costs and lack of labor result in curtailment of development work. Ore reserves become depleted and grade of ore sent to the mill is lowered by the inclusion of more and more poor rock in order to keep up tonnage. For a short time there may be a deceptive lowering of cost because of lack of development work. But the secret of successfully mining the narrow veins of the district lies in keeping up a good-sized reserve by adhering to a definite program of development. Under present conditions this can be done only by the richest companies. The detailed costs as reported by the general manager for 1917 were:

### GENERAL EXPENSE.

|  | Per ton |
|---|---|
| Administration, salaries | $0.2824 |
| New York office expense | .0245 |
| Mine expense | .0624 |
| Taxes | .0914 |
| Fire insurance | .0221 |
| Casualty insurance | .1741 |
| Interest | .0097 |
| Total general expense | $0.6666 |

### MINING.

| Labor | $2.7422 |
|---|---|
| Power | .5480 |
| Supplies | .8523 |
| Pump labor | .1694 |
| Pump supplies | .0052 |
| Total | $4.3171 |

### MILLING.

| Labor | $0.3142 |
|---|---|
| Power | .1634 |
| Supplies | .1147 |
| Bullion, freight, refining | .0189 |
| Concentrates, freight and treatment | .2991 |
| Total | $0.9103 |

### NEW SHAFT.

| Labor | $0.1198 |
|---|---|
| Power | .0191 |
| Supplies | .0246 |
| Total | $0.1635 |

REPAIRS.
(Repairs to mill, telephone line, pumps, and buildings) _____ $0.0692

IMPROVEMENTS.
Mine equipment _____ $0.1197
Tailings, dam _____ .0121
Mine phone installation_____ .0005
Foreman's dwelling supplies_____ .0006

Total _____ $0.1329

The total of costs normally chargeable to the one year's operation,
according to the data furnished, is thus $6.26 per ton. Besides this
there was a considerable outlay for new property.—*C. A. Logan.*]

**Buckeye Claim.** Owner, Cold Springs Quartz and Gravel Mining
Company, Nevada City. Under option to A. Hoge, Nevada City.

> Location: Nevada City District, Sec. 4, T. 16 N., R. 9 E., 3 miles northeast of
> Nevada City.
> Bibliography: Cal. State Min. Bur. Rept. XIII, page 237.

The Buckeye quartz claim, together with the Cold Springs gravel
properties which are owned by the same parties, total 1000 acres.

The Buckeye vein, outcropping in the Calaveras slates about 500'
east of the granodiorite contact, strikes nearly north and dips 45° to 60°
east. A tunnel was driven northward from the bank of Willow Valley
Creek, a distance of 500' on the vein, and two ore shoots were stoped
120' to the surface. An inclined shaft was also put down on the vein
to a depth of 210', and drifts were run 150' north and 300' south. The
vein varies in width from 1' to 4' and can be traced for a distance of
1500' on the surface until it disappears under the andesite to the north.
The property was idle when visited.

**Buckeye Extension.** (See Fountain Head Mining Company.)

**Buena Vista Claim.** Owner, S. P. Dorsey, Grass Valley.

> Location: Grass Valley District, Secs. 2 and 11, T. 15 N., R. 8 E., 2½ miles
> south of Grass Valley. Elevation 2300'.

This claim adjoins the Allison Ranch mine on the south and covers
about 1500' along a vein belonging to the Omaha-Allison Ranch vein
system. The vein dips to the west and has been traced at intervals for
several hundred feet by shallow shafts and open cuts, the only work
that has been done.

**Bullion Consolidated Mine.** Owners, Bullion Consolidated Gold
Mining Company; John Martin and E. de Sabla, Alaska Commercial
Building, San Francisco.

> Location: Grass Valley District, Secs. 1 and 2, T. 15 N., R. 8 E., 2 miles south
> of Grass Valley. Elevation 2500'.
> Bibliography: Cal. State Min. Bur. Rept. XIII, page 237.

The Bullion vein lies 3000' east of the west-dipping Omaha-Allison
Ranch vein, near the eastern contact of the granodiorite and has an

easterly dip of 33°, contrary to the general westerly dip of the Osborne Hill and Omaha vein systems. The Union Jack, Smuggler, Bullion and La Bruja together with 40 acres of agricultural land comprise the Bullion holdings which control 5000' along the strike of the vein. The quartz vein outcrops in granodiorite which forms both walls to a depth of 200'; below this level it passes into diabase; it varies in width from 1 to 5 feet and the ore carries free gold, pyrite and a considerable amount of galena. The sulphides average 4% and are valued at $75 per ton.

The mine was worked in the 60's and since that time has been developed to a depth of 1500' by means of an incline shaft which, start-ing in the hanging wall, encounters the vein at a depth of 300 feet from which point it follows the vein. Levels have been run as follows: 300' and 400' levels both driven 150' north and 150' south of the shaft; 600' level driven 150' north; 700' level driven 300' north and 300' south; 900' level driven 250' south; 1000' level driven 500' north; 1100' level driven 700' north and the 1500' level driven 750' north and south of the shaft. A crosscut was also driven from a point 200' south of the shaft on the 900' level a distance of 150' west into the footwall of the vein. This development work opened a number of ore shoots and ore has been stoped therefrom in the following places; on the 300' level from a shoot 100' in length; on the 700' level north of the shaft from a shoot 150' in length; on the 1100' level 600' north of the shaft from a shoot 100' in length; and on the 1500' level, 1700' north of the shaft, where considerable ore is said to be still in place. The mine was last worked in 1906 and the equipment which consists of boilers, motors, a 14" steam driven Cornish pump, and 10-stamp mill, would in all proba-bility have to be replaced if the mine is reopened. The adjoining mine to the south, the Alaska, is supposed to be on the southern extension of the Bullion vein.

The Diamond mine now owned by the Bullion Company is located on patented agricultural ground about 2500' northwest of the Bullion shaft. The Big Diamond vein strikes N. 20° W. and, like the Galena and Bullion vein, of which it is in all probability the north extension, dips 45° east. The Little Diamond, an east-west vein dipping 48° south, crosses the Big Diamond vein. These veins outcrop in granodiorite and have been developed by a vertical shaft and an inclined shaft 290' deep. A 1000' tunnel was also driven to the vein. The deepest work-ings below the outcrop are about 125' but 1200' of exploratory drifts have been driven on the veins which vary in width from 18" to 3', and in early days a considerable amount of $20 ore was stoped.

**Cabin Flat Mine.** Owner, Golden Center of Grass Valley Mining Company, Grass Valley.

Location: Grass Valley District, Sec. 27, T. 16 N., R. 8 E.; adjoins townsite of Grass Valley on the west.

This mine, which lies north of the Gold Hill mine, has not been worked for many years. The vein is the northern continuation of one of the veins worked in the Gold Hill, but it is small, averaging only 12″ in width. It dips to the east and the only development work consists of a few shallow shafts.

**Cadmus.** (See Champion.)

Bibliography: Cal. State Min. Bur. Rept. XIII, page 237.

**California.** (See under Copper.)

**California Mine** (Pittsburg). (See also **West Point** and **California Mines Co.**) Last operated under bond by California King Company (King C. Gillett).

Location: Rough and Ready Mining District, 3½ to 4 miles from Grass Valley.

The following notes are from a report on the mine by the author, in 1918:

The property comprises two claims, the Pittsburg, and Pittsburg North Extension, both patented, covering 3000 feet along the strike of the veins.

Previous to work by above company, the following development had been done:

At a point practically on the lode line, and 30 feet south of the common end line, a shaft had been sunk 250 feet and drifts driven as follows:

On the  50-foot level, drifted 100 feet southwest and 100 feet northeast.
On the 100-foot level, drifted 100 feet southwest and  60 feet northeast.
On the 150-foot level, drifted 110 feet southwest and  80 feet northeast.
On the 200-foot level, drifted 100 feet southwest and 140 feet northeast.
On the 250-foot level, drifted 360 feet southwest and 400 feet northeast.

Above 150 ft. level, the vein was stoped 75 ft. on each side of shaft. On 200 ft. level just north of shaft, a stope 40 ft. long was carried up 40 ft.; and 20 ft. southwest of shaft a stope 40 ft. long was carried up 30 ft. No stoping was done between the 200 ft. and 250 ft. levels.

The California King Company sank a new shaft 600 ft. deep at a point 150 ft. southwest of old shaft. At a depth of 150 ft. below the old 250 ft. level, drifts were driven 60 ft. southwest and 90 ft. northeast. No drifts were driven below the 400 ft. level.

The vein is a well defined fissure vein in a complex of gabbro-diorite and amphibolite. It strikes N. 30° E., dips 65° to 80° NW.; and varies in width from a few inches to 4 feet. The vein filling consists of small auriferous quartz stringers a few inches wide, and barren altered wall rock. The foot-wall is well defined, with several inches of gouge, to the bottom of shaft. The auriferous quartz follows the hanging wall.

Oxidation and mechanical concentration have enriched the vein to a depth of about 200 feet. The gold and auriferous sulphides are evidently confined to the small quartz stringers. No well defined ore shoots were developed. The rich bunches which yielded a few tons each in the upper levels are less numerous with increasing depth. Some of the narrow quartz stringers give high assays.

The California King Company apparently quit work because the ore-bodies blocked out between the two shafts and the 250 ft. and 400 ft. levels did not seem to be of high enough grade to pay expenses. The characteristics of the veins in this district are held to be entirely different from those of persistent veins in the Grass Valley district. However, the veins in the Rough and Ready district have never been explored at such depths, as have been attained on the North Star and Empire properties.

**California Consolidated Group** (Gold Tunnel). Owner, G. G. Allen Estate; care of Mrs. Margaret Gilbert, Nevada City.

Location: Nevada City District, Sec. 12, T. 16 N., R. 8 E., in Townsite of Nevada City. Elevation 2500'.
Bibliography: Cal. State Min. Bur. Rept. XIII, page 246.

The property of the California Consolidated Company, consists of the California, and the Gold Tunnel mines; the latter was the first quartz mine to be located in the Nevada City district and was worked from 1850 to 1875 with but few interruptions. It was again worked in 1895 but since that date has been idle for the greater part of the time.

The Reward-Gold Tunnel-Oustomah vein can be traced from the Reward shaft northward a distance of 7000 feet to the Oustomah mine where it disappears under the lava capping. The lode occurs in the granodiorite and has an average dip of about 30° east. It varies in width from a few inches to 4' and has been developed on both the Gold Tunnel and California claims by means of tunnels driven north and south from the banks of Deer Creek and by inclined shafts. The workings on the California reached a depth of 700' on the vein and those on the Gold Tunnel 300' below the tunnel level. Considerable lateral development work was done from the inclined shafts. The production of these properties can not be definitely ascertained but it is estimated to have been nearly $1,000,000. The mine was idle in 1914.

**California Gold and Copper Company.** (See under Copper.)

**Calumet and California Mining Company,** Geo. C. Morrison, president; Elbert E. Boyd, secretary, Calumet, Michigan.

This company was reorganized from the Fairview Mining Company, in 1914, for the purpose of continuing work at the Fairview Mine (which see). In 1916 work was kept going by assessments. A statement issued by the company in September, 1916, shows that out of $14,067 collected, only $6,624 was spent on actual exploration. The money was exhausted in July, 1917, and the mine was abandoned, apparently without any recent production. The mine, 20-stamp mill and equipment have been sold to the Sierra Asbestos Company, and the mill was being used in September, 1918, to crush chrysotile asbestos.

**Cambridge.**   (See Union Hill.)

**Canada Hill Consolidated Group.**   (Charonatte).   Owner, Ralph Gaylord, Nevada City.

Location: Nevada City District, Sec. 17, T. 16 N., R. 8 E., 2 miles east of
    Nevada City.   Elevation 2700′.
Bibliography: Cal. State Min. Bur. Rept. XIII, page 238.   U. S. Geol. Survey
    Folio 27, maps.

The claims composing this property are: the North Grant controlling 1500′ along the Grant Canada veins; the Canada Hill Extension claim controlling 1400′ on the Grant and 1100′ on the Canada Hill vein; and the Independence claim controlling 1440′ of the north extension claims Greenman ledge. The St. Louis vein crosses this group of claims, passing through the side lines of the Grant and Independence claims. The Canada Hill vein occurs in the granodiorite and the northern end is faulted 150′ west by the St. Louis cross vein. It has been developed by an inclined shaft on the vein 1500′ in depth and 10,000′ of drifts driven therefrom. The mine was worked from 1854–1863 and again from 1879 to 1887; during the later period 19,810 tons of an average value of $18 per ton were produced. The mine was again operated in 1908–1910 when a few hundred feet of development work was done; since that time it has been idle.

The Grant vein, lying parallel and east of the Canada Hill averages 16″ in width, dips at an angle of 45° to the east and crosses the contact of the metamorphosed Calaveras slates and the granodiorite. A 300′ incline shaft was sunk on this vein near the contact. The Greenman ledge which has a north strike and dips 15° to 30° outcrops 450′ west of the Canada Hill vein which it intersects at a depth of 250′ on the dip. The Canada Hill is not faulted by the Greenman vein which is said to continue beyond on its regular dip after intersecting the Canada Hill. The development work consists of a tunnel 600′ in length on the vein, 80′ below the surface and a 200′ incline on the vein. Ore taken from the workings is said to have yielded from $35 to $60 per ton.

The St. Louis vein, which can be traced on the surface for nearly 7000′, strikes N. 65° E. and dips 70 to 80° N. This well-defined fissure vein, including the altered granodiorite which forms both walls, varies in width from 1 to 12 feet. The ore is of much lower grade than that found in the north and south veins. The drain tunnel which taps the

Canada Hill shaft was driven on the St. Louis vein for a distance of 600' and this is the only work so far done on the vein, with the exception of some short tunnels on the adjoining Alpine ground.

The total production of the various veins on the Canada Hill property is estimated at $500,000. The equipment consists of a hoist, three pumps and the necessary buildings. The mine can be reopened with the present equipment.

**Carlisle.** (See under Copper.)

**Carter Group.** Owners, Haggin & Hearst, San Francisco.

Location: Nevada City Mining District.
Bibliography: U. S. Geol. Survey, W. Lindgren. Prof. Paper 73, pages 125–132. 17th Annual Report, pt. II, pages 1–262, 1896. U. S. Geol. Survey Folios 18 and 29, Smartsville and Nevada City.

This group of quartz claims at Missouri Bar on the South Yuba River, was purchased by Stuart J. Rawlins, representing the Hearst Estate of San Francisco, in the summer of 1915. The holdings contain a large ledge of milling ore, of which the greater portion is free milling.

**Cassidy Mine.** (Linden.) Owner, Empire Mines and Investment Company.

Location: Grass Valley District, Sec. 35, T. 16 N., R. 8 E., 1 mile southeast of Grass Valley. Elevation 2600'.
Bibliography: U. S. Geol. Survey Folio 29.

The Cassidy and Linden veins belong to the great west-dipping Empire-Osborne system of fissures. The Linden vein lies from 600' to 800' west of the Empire-Rich Hill vein and parallel to it. The Cassidy vein outcrops from 300' to 600' west of the Linden and has the same general north strike. The property consists of two patented claims, the Cassidy (O'Connor and Gilroy) and the Linden, a total area of fifty acres. These claims cover 1800' along the Cassidy and 2500' along the Linden vein. Both veins were worked in early days to a depth of 50' by shallow shafts and cuts. In 1910 the mine was operated and in August, 1914, the property was under bond and being unwatered.

The Cassidy vein which on the surface crosses the contact of the diabase and granodiorite, has been developed to a depth of 400' by an inclined shaft sunk at an angle of 32° in the hanging wall. Drifts were run at 250' and 350' (177' vert.).

The mine is equipped with an Ingersoll-Rand Imperial type compressor of 200 cubic feet capacity driven by a 100-horsepower, 400-volt G. E. induction motor; a Laidlaw-Dunn-Gordon 12" x 12" compressor, 150 cu. ft. cap., a 50-horsepower G. E. motor, a small double driven 15-horsepower hoist, and a blacksmith shop completely equipped. All the machinery is in excellent condition and housed in galvanized iron buildings.

Taken under bond by the Linden Mining Company in May, 1915, and electric pumps installed; later purchased by Empire Mines and Investment Company.

**Cedar Claim.** Owner, Cedar Mining Company; care of Nickerson Estate, Auburn.

Location: Secs. 20 and 29, T. 14 N., R. 8 E., 14 miles south of Grass Valley. Elevation 1200'.
Bibliography: Cal. State Min. Bur. Repts. XII, page 186; XIII, page 238.

There is a strong vein, of which the Cedar Mining Company controls about 5000', which strikes north crossing south Wolf Creek near its junction with Bear River. The lode occurs in diabase near its contact with a narrow belt of Calaveras slates. The vein filling is composed of quartz and calcite, carrying a heavy percentage of auriferous chalcopyrite. Only superficial development work has been done. The mine was idle in 1914.

**Celia Mine.** Owners, Fritz Meister Estate and Mrs. R. A. Keenan, Washington, Nevada County, California.

Location: Washington District, Sec. 16, T. 17 N., R. 11 E., 3 miles southeast of Washington by fair wagon road. Elevation 4500'.

This mine consists of the Alta and Burghman, patented lode claims, and the Fritz Meister patented placer mine of 40 acres, making a total area of 80 acres. The property adjoins the famous Omega hydraulic mine on the east. The lode claims cover 3000' along the Celia vein which outcrops in the Calaveras slates and strikes north. It belongs to the same vein system as the Cooley and Washington veins which have been developed on the South Yuba River one mile north, at an elevation 1500' lower than that of the Celia. The property has only been superficially developed, and no work has been done for a number of years.

**Centennial Claim.** (See Sultana.) Owner, Sultana Gold Mining Company, Crocker Building, San Francisco.

Location: Grass Valley District, Sec. 1, T. 15 N., R. 8 E., 2 miles southeast of Grass Valley. Elevation 2600'.

**Central and Gray Eagle Group.** (See Yuba Consolidated.) Owner, Yuba Consolidated Gold Mining Company, San Francisco.

Location: Washington District, 5 miles northeast of Washington. Elevation 4000'.
Bibliography: Cal. State Min. Bur. Rept. XIII, page 238.

Idle in 1918.

**Central-North Star Mine.** (See North Star.) Owner, North Star Mines Company.

Location: Grass Valley District, Sec. 3, T. 15 N., R. 8 E., 2 miles south of Grass Valley. Elevation 2480'.
Bibliography: Cal. State Min. Bur. Repts. XI, page 277; XII, page 117; XIII, page 238.

**Central South Yuba Mine.** Owner, Queen Regent Merger Mines Company.

Location: Nevada City Mining District.
Bibliography: U. S. Geol. Survey, W. Lindgren, Prof. Paper 73, pages 125–132. U. S. Geol. Survey 17th Annual Report, pt. I, pages 1–262, 1896. U. S. Geol. Survey Folios 18 and 29, Smartsville and Nevada City.

Property is situated a few miles below Grass Valley. In October, 1915, the first shipment of ore, about thirty tons, was consigned to the Selby smelter. It assayed around 12% copper, 5% zinc, $12 gold and $6 silver. The ore body is 4' wide and has been drifted on for 120' from the shaft on the 100' level. As the mine is several miles from the railroad, only the best ore has been mined in the past. In December, 1915, the Queen Regent Merger Mines Company purchased the property and

Photo No. 17. Headframe and plant at the Champion shaft near Nevada City.
Photo by C. A. Logan.

arranged to install a reduction plant. The plant will be of an experimental type and is expected to make an extraction approximating 95% of the assay value.

**Champion Group.** Includes Providence, Home (Cadmus) Merrifield, Spanish, Wyoming and Nevada City mines. Owner, North Star Mines Company, 22 William street, New York; George B. Agnew, president, New York; A. D. Foote, manager; A. B. Foote, superintendent.

Location: Nevada City District, Secs. 2, 11, 12, 13, and 14, T. 16 N., R. 8 E., 1½ miles west of Nevada City. Elevation 2300'.
Bibliography: Cal. State Min. Bur. Repts. VI, pt. II, page 47; VIII, page 418; XI, page 290; XII, page 198; XIII, page 260. U. S. Geol Survey 17th Annual Report, pt. II, Gold Quartz Veins of Nevada City and Grass Valley, Lindgren.

The Champion property consists of a consolidation of a number of mines and claims, containing about 440 acres and controlling 8000 feet along the Ural and Merrifield veins. The following claims now compose the Champion group; the Bavaria, Bayard Taylor, Champion, Clima, Deer Creek (Cadmus) East Home, Home North, Home South, Mary Ann, Miller, Nevada (Merrifield), Nevada City Extension, New Years, New Years Extension, North Wyoming, Phillip, Providence, Schmidt (Nevada City), Soggs, Spanish, Ural (Cornish), Ural (relocation), West Providence, West Providence Extension, West Wyoming, Wyoming, Graves Placer mine, and Swartz Placer mine.

The Providence, Champion, Nevada City, Wyoming and Home have in the past been the principal producing mines, and the combined

Photo No. 18.   Providence Mine, surface plant and waste dump.   This shaft is on the south bank of Deer Creek, just opposite the Champion.   Photo by C. A. Logan.

production is variously reported to have been from $8,000,000 to $20,000,000.

*Champion Mine.*   The Merrifield and Ural, together with the Wyoming and other minor veins, comprising one of the most important vein systems of the Nevada City District, are all controlled by the Champion holdings.

The Merrifield vein which varies in strike from N. to N. 20° W. and has an average dip of 35° E., is one of the longest and most persistent veins in the Nevada City district.   It can be traced from a point 3000' south of the Providence shaft to the Mount Auburn mine, a total distance of 11,000 feet.   South of the Providence shaft the Merrifield vein at the surface occurs near and at the contact of the slates and granodiorite. North of the Providence shaft the vein occurs wholly in the granodiorite. The fissures are the result of intense dynamic stresses which, according to Lindgren, resulted in a movement of the granodiorite hanging wall upward a distance of 1200 ft. As a result of the intense movement, there is generally a zone of crushed and altered material reaching a maximum width of 30 ft. The Merrifield quartz veins occurring in this altered zone vary in width from one to ten feet. Of the wider veins from four to six feet is solid quartz. In some cases the orebodies occur along parallel planes separated by a few feet of altered country rock, while in others they are lenticular in form. The ore is a milky white quartz carrying an average of 6% of sulphides, consisting of pyrite, chalcopyrite, galena and sphalerite. 'Specimen ore' rarely occurs.

The *Ural* vein, which lies parallel and 500' west of the Merrifield, follows the contact of the granodiorite and contact metamorphosed Calaveras formation. It can be traced from the Champion shaft, N. 15° W. for a distance of 3500', to within a short distance of the 'New' Nevada City shaft. At this point the vein takes a sharp bend to the westward (N. 70° W.) following the contact and can be traced to the Coan mine a distance of 4000', making a total length of 7500 ft. From the point mentioned, a branch vein continues northward for a distance of 1000' into the granodiorite, but appears to die out before intersecting the northern extension of the Merrifield vein. The Ural vein has an average dip of 35° E. and was worked in crosscuts from the 600', 1200' and 1800' levels. On the 600' level a drift was driven south from the crosscut on the Merrifield vein. This drift shows that 400' south of the crosscut the vein enters an altered diabase and then splits, the east branch continuing south and then east in the diabase, while the west branch taking first a westerly and then a southerly course, finally enters an unaltered black slate. In the Providence mine the main ore body lies north of the point where the vein enters the diabase and was about 150' in length with an average width of from 2½ to 3 feet. The vein carries a black gouge on the foot-wall and several feet of altered granodiorite on the hanging wall. The sulphides are pyrite, chalcopyrite, galena and zinc blende amounting to from 5 to 8 per cent, and an occurrence of telluride of silver and lead is reported.

The physical and mineralogical characteristics of the Ural vein, as shown by the Champion workings, are similar to those in the Providence ground. The Wyoming vein, which lies in the footwall of the Ural vein, dips to the east and joins the latter in depth. It seems probable that this vein in its northern extension is the one that intersects the Ural vein in the vicinity of the 'Old' Nevada City shaft.

The northern continuation of the Ural vein in the Nevada City (Gold Hill) property has been extensively worked since 1879, and in 1914 most of the ore being produced was coming from this source.   The vein averages about 2' in width, reaching at times a maximum of 12 feet. As in the case of the Ural vein in the Champion-Providence territory the lode in the Nevada City ground shows the characteristic evidence of intense movement and alteration.   The sulphides, however, average only 2½ per cent with a value of $150 per ton.

For a detailed geologic and mineralogic description of the Ural and Merrifield veins, see U. S. Geol. Survey, 17th Ann. Report, part II, "Gold Quartz veins of Nevada City and Grass Valley," by Lindgren; and U. S. Geol. Survey, Nevada City Special Folio.

The *Providence Mine,* located in 1858 as one of the earliest claims of the group, was first operated from 1861 to 1867.   Owing to difficulties encountered in treating the concentrates the mine was not worked successfully until after the adoption of the chlorination process about 1870.   In 1886 the shaft had reached a depth of 1100' and eleven drifts, the longest of which was 3600', had been driven on the Merrifield vein. A crosscut 547' in length had also been run westward to the Ural vein from the 600' level of the Providence shaft.   After being worked at intervals by the owners and tributers the mine was closed in 1891, but was again reopened the following year.   In 1894 apex litigation was instituted between the Providence and Champion and this was finally settled in 1902 by the purchase of the Providence property by the Champion company.   The Providence had been developed to a depth of 1750', and was credited with a production of $5,000,000 prior to its acquisition by the Champion.

The Champion property, located in 1851 as the New Years, was not opened as early as the adjoining Providence mine and in 1888 the inclined shaft had reached a depth of only 300' (180' vertical) and only 350' of drifts had been driven.   In 1892 a consolidation with the adjoining Merrified and 8 other claims had been effected and the development work consisted of a 3000' drain tunnel driven from Wolf Creek on the vein and intersecting the shaft at a depth of 600'; an inclined shaft on the vein 1000' in depth and the ten levels had reached a maximum length of 1000' north of the shaft.   Although compelled to suspend operations at intervals during the period from 1892 to 1902, owing to the Providence-Champion legal war, the Champion shaft had been sunk to a depth of 2400' on the vein and a new shaft on the Merrifield vein had reached a depth of 900 feet.   As has been stated before, the Providence property was purchased by the Champion in 1902. In 1905, however, the Champion was again involved in litigation with the Home (Cadmus) mine, and this suit was also settled in the early part of 1907 by the purchase of the Home holdings by the Champion company. The Home and Cadmus claims lying west of the Providence were developed by short tunnels and shallow shafts until about 1896, when active work began, and in 1902 the Home shaft had reached a depth of 700', and a 30-stamp mill had been installed.

The Nevada City or Gold Hill mine in which the Northern extension of the Ural vein had been extensively worked from 1880 to 1895 was included in the Champion holdings by purchase, after suffering from legal complications. Two incline shafts were sunk on the vein, one on the Schmidt claim near the southern end line 1000' deep and the 'new' shaft on the Nevada City claim 1100' northwest of the old shaft. From 1907 to 1911, the Champion mines experienced the usual vicissitudes of fortune that generally follow protracted litigation, when money is used for attorney fees, rather than for systematic development work. In June, 1908, work was suspended, but a few months later tributers were working, and in 1909 the mine was bonded. Later the bond was forfeited, which led to more law suits. In 1911 the North Star Mines Company bonded the property, and after extensive exploration work the Champion group was purchased by the North Star Mines Company. The work so far accomplished by this company consists of unwatering the Champion shaft to the bottom or 2400' level; reopening and driving the 1000' level 2500' north on the Ural vein to the Nevada City ore shoot; driving the 2400' drift north 3000'; crosscutting on the 1600' level connecting the Ural and Merrifield veins, and a drift which was driven a distance 2500' north of the crosscut on the Merrified vein. Most of the ore being produced in 1914-15 was obtained from the ore bodies of the Nevada City mine.

Owing to the fragmentary records of the gold output of the different properties, it is impossible to accurately ascertain the total production of the consolidated properties, which as previously stated has been variously estimated over the wide range of from $8,000,000 to $20,-000,000. The latter figure is undoubtedly too large. From authentic records, the production of the Champion Company from 1893 to 1913, has been $2,864,528 from 508,910 tons of ore. This total, however, does not include the production of the Providence or Home mines prior to their purchase by the Champion. According to Lindgren the production of the Providence mine from the Ural and Merrifield veins prior to 1896 was estimated at $5,000,000, but this figure was considered excessive. From the data available, it seems probable that the combined production of the Ural and Merrified veins has been between eight and nine million dollars.

### New Developments and Additions to Plant.

[Since the foregoing was written, extensive changes have been made above and below ground on this group.

The Providence shaft has been sunk to 2700 ft. and enlarged to three compartments with 18 feet in the clear, and with a water hoist in the third compartment. It follows the vein all the way, dipping 38° to 40°. Thirty-pound rails have been laid in this shaft. A new and larger headframe has been built over it. A fine new compressor is in use. It is a Laidlaw-Dunn-Gordon 20″ x 12″ x 24″, 1200 cu. ft. capacity free air per minute, 225 horsepower, two-stage, variable volume, direct-connected. A 100-horsepower double-drum electric hoist equipped with two-ton skips has been installed. The Providence shaft is now the main hoisting shaft. From this shaft ore is delivered to the mill across Deer Creek by a rope tramway carrying buckets of 550 pounds capacity over a 450 ft. span. This tram is automatic except as regards filling, and has been found capable of delivering a bucket of ore every 55 seconds, or 18 tons an hour.

The Champion shaft has also been sunk to 2700 feet. From the 2400-ft. level of the Champion a drift has been driven to the north toward the Nevada City vein and a raise put through to connect with the Nevada City shaft at the end of the north drift on the 1000 ft. level of the Champion. The Nevada City shaft is now a main airway, and the Champion shaft is used largely for lowering timbers, etc. It also contains a Cornish pump. The Champion and Providence workings are connected on the 600′, 1200′, 1600′, 1750′ and 2700′ levels.

The principal stoping now (September, 1918) is being done from the 2700 ft. south drift of the Providence, where the shoot has a stope length of 300′ to 400′, and is 2′ to 10′ thick. The ore shoot in the Providence is a remarkably strong and persistent one, having been followed from the surface to the 2700 ft. level. A winze is being put down from the 2700 ft. level, and a drift has been started at 2800 feet. From the 1600 ft. Providence north drift a winze is being sunk, and a raise will be put up from the 2700 ft. north drift to connect with it. A drift started north from this winze at 1800 feet is already showing good ore.

In the Champion some stoping has been done from the 2700 ft. drifts and a little from the 2400′. From the 'C' level (about 1600 feet) a small stope was carried up through the 1000 ft. north drift to the 700 ft. level on the Nevada City ore shoot. The Champion ore shoot was bottomed at about 1200 feet.

*Pumping.* On the 2700 ft. level of the Providence there is a three-throw plunger pump using 36 to 40 horsepower and handling 200 gallons of water a minute against 530 ft. head to the 1650 ft. level. The main pumping station is at the latter level. Here a 5-throw plunger pump, 225 horsepower, handles 550 gallons a minute against 840 ft. head to the drain tunnel, which is about 100 ft. below the collar of the shaft on the incline. The workings of the Providence are under the creek and a good deal of water comes in near the 600 ft. level.

The Cornish pump in the Champion shaft handles from 60 to 80 gallons of water a minute from the 1000 ft. level. Another Cornish pump at the Cadmus shaft handles 30 gallons a minute.

*Ore and mining conditions.* The veins in these properties are entirely different from those at Grass Valley, and different methods of timbering and mining are required. The veins are generally not less than 2 feet thick, and dip at an average angle of 35 degrees to 40 degrees, so that less shoveling is necessary than at Grass Valley, and less waste has to be handled. There is no specimen ore. The concentrates are said to average about 6% of the ore and carry about 30% of the value. The ground is heavy and cost of timbering is high. Modified square setting is used. The life of timbers is said to be not over a year and a half. Long working drifts have to be watched and the ground taken frequently, and caution is desirable in the stopes. According to Weed* the developments at the Champion group up to 1917 had not been satisfactory, the average yield not paying for operation in 1916. At present, however, the superintendent expresses satisfaction at the grade of ore, and it is evident that conditions have improved greatly since stoping in the deeper levels of the Providence began.

*Milling.* The stamp mill, which has been remodelled, contains forty 1250 lb. stamps. Ore entering the mill from the tramway goes over $1\frac{1}{4}''$ grizzlies, oversize being broken. Thence it goes to stamps which drop $6\frac{1}{2}''$ to 7", 106 times a minute. Outside amalgamation only is used, and the pulp flows over 6 ft. amalgamating plates to Frenier sand pumps, which send it to hydraulic classifiers. The slime goes to thickeners; the sand is concentrated on three double deck and two single deck Deister tables. The concentrate is dewatered, after which it is ground in a tube mill. The middling is sent to two Union Vanners. Tailing goes from the tables to 7 sand vats. From the tube mill the ground concentrate flows over a 6 ft. amalgamating plate, fines going to cyanide and portion over 200 mesh returning to tube mill for regrinding.

The slime from thickener is treated in 6 Pachuca Agitators. The concentrate is treated in charges in Pachuca Agitators and pumped into the slimes, which are filtered on a 10' x 12' Oliver Filter. Merrill presses use common building paper on the frames; this is found to be a satisfactory substitute for cloth.

---

*Weed, W. H., The Mines Handbook, Vol. XIII, 1918.

About 80% of the gold saved is recovered by cyanidation and 20% by outside amalgamation. In September, 1918, five stamps were being used about ¾ of the time crushing chromite ore from nearby properties.

Photo No. 19.    Mill and cyanide plant of the Champion-Providence Group near Nevada City.    Photo by C. A. Logan.

The capacity of the mill is upward of 5000 tons a month when in full operation. The stamp mill, etc., used 100 horsepower; the tube mill 25 horsepower and the cyanide plant 25 horsepower. The concentrate varies widely in value, ranging from $50 to $120 a ton. Flotation tests have recently been made at the Champion, and it is probable that this system will be installed later, when business conditions become normal.

Since the Champion-Providence group is the only active mine in the district it is thought desirable to show the latest obtainable detailed costs, which are for 1917, as furnished by Roy Stapler, Assistant Superintendent.

**Mining Cost.**

| | Per ton mined |
|---|---|
| Breaking rock (operating) | $0 02 |
| Breaking rock (development) | 25 |
| Timber (operating) | 85 |
| Timber (development) | 06 |
| Machinists and mechanics | 04 |
| Shoveling | 32 |
| Tools | 13 |
| Drills | 29 |
| Repairs (shafts, drifts, etc.) | 27 |
| Tramming (operating) | 16 |
| Tramming (development) | 22 |
| Hoisting | 3 |
| Surface tram (charged to underground mining) | 07 |
| Pumping (500,000 gallons in 24 hours) | 38 |
| Bosses (including officials, assayer, etc.) | 23 |
| Miscellaneous (electricity, moving dump tracks, etc.) | 10 |
| Total mining cost per ton mined | $4 42 |

**Milling Cost.**

| | Per ton |
|---|---|
| Milling and concentrating | $0 50 |
| Cyaniding | 68 |
| Tailings disposal | 13 |
| General expense, accidents and insurance | 18 |
| | $1 39 |
| | 4 42 |
| Total cost, mining, milling, general | $5 81 |

*—C. A. Logan.*]

**Charonatte.**  (See Canada Hill.)

**Cincinnati Hill Claim.**  (See North Star.)  Owner, North Star Mines Company.

Location: Grass Valley District, Secs. 28 and 33, T. 16 N., R. 8 E., ½ mile west of Grass Valley.  Elevation 2500'.

**Coan Mine.**  Owner, T. N. Coan, Nevada City, California.

Location: Nevada City Mining District, Sec. 3, T. 16 N., R. 8 E., 3 miles west of Nevada City.  Elevation 2400'.
Bibliography: Cal. State Min. Bur. Rept. XIII, page 239.  U. S. Geol. Survey Folio 29, Nevada City Special.

The Coan mine consists of 21 acres of patented railroad land, covering 1470' along the probable westerly extension of the Ural-Champion vein.  The course of the vein at this point is N. 75° W. and it dips 45° N.  The vein occurs in the diorite and aplite near their contact with the metamorphosed Calaveras slates.  The mine was worked in early days to a depth of 100 feet. It was reopened in 1911 by W. A. Hilliard and associates, and the inclined shaft was sunk to a depth of 500' and drifts were driven on the vein as follows: 100' level 340 ft. east and 225 ft. west; 200' level 165 ft. east and 100 ft. west; 500' level 1000 ft. east and 200 ft. west. The vein as thus developed varies in width from 1' to 20' with an average of about 3 feet. There is also a so-called 'back ledge' upon which some work has been done. The quartz carries free gold, pyrite, chalcopyrite, galena, and sphalerite. Sulphides vary in value from $40 to $100 per ton. The length of the ore shoot developed was about 165', but only a small amount of ore has been stoped. The mine is equipped with a modern shaft house and hoist, a four-drill electrically-driven Sullivan compressor, air and Cornish pumps, air drills and stopers, and a 10-stamp mill. The mine was last operated in 1913 when 19 men were employed under the management of W. A. Hilliard of Grass Valley.

**Coe Mine.** Owner, formerly Coe Quartz Mining Company; sold to State for taxes.

Location: Grass Valley District, Sec. 27, T. 16 N., R. 8 E., ½ mile north of Grass Valley. Elevation 2520'.

The Coe mine, which consists of one claim of irregular shape 2000' by 200', was worked in early days to a depth of 550' by means of an inclined shaft and is said to have produced $500,000. After being closed for 25 years, it was again reopened in the spring of 1900. The mine was completely equipped, the shaft was reopened and a 20-stamp mill was erected. After sinking the shaft to a depth of 1150' and driving 3000' of drifts, it was closed in October, 1904, since which time the equipment has been removed and the mine has remained in idleness. The vein, which strikes N. 80° W. and dips 65° N., occurs in serpentine near its contact with the Calaveras slate. The lode varies in width from 1' to 3' of solid quartz and the sulphide consists of 1% of pyrite and galena.

The old hoisting plant was destroyed by fire in the early part of 1916.

**Colling Mine.**

Location: Rough and Ready Mining District.
Bibliography: U. S. Geol. Survey, W. Lindgren, Prof. Paper 73, pages 120–124. U. S. Geol. Survey Folio 18, Smartsville.

After having been closed for two years, this mine was reopened in January, 1915. Three small veins crossing the main ledge were said to show good ore. Developed by a three-compartment shaft 132' in depth.

**Columbia Group.** Owner, Columbia Consolidated Mines Company; H. L. Hughson, president, Mills Building, San Francisco; John Morgan, secretary, Citizens Bank, Nevada City.

Location: Washington District, Sec. 4, T. 17 N., R. 11 E., 2½ miles east of Washington. Elevation 3000'–4500'.
Bibliography: U. S. Geol. Survey Folio 66, Colfax.

There are four patented claims in this group, Canyon Creek, Canyon Creek Extension, Waldeck, and Columbia, and one location, the Carlotta. These claims comprise an area of about 80 acres and cover 3000' along the strike of the veins. The four veins developed on the property were discovered in early days and were worked by means of tunnels from the west bank of Canyon Creek, which cuts through the veins at a point several hundred feet above its junction with the South Yuba River.

Photo No. 20. Outcrop of the Columbia Lode near Washington.

The four veins on the Columbia ground dip nearly vertical and all strike N. 20° W. in the Calaveras (Blue Canyon) slate and quartzite.

Most of the development work so far accomplished has been done on vein No. 3, which is said to vary in width from 5' to 20'; it is the most westerly of the four. Two hundred feet east of vein No. 3 lies No. 4 which contains about 5' of quartz; vein No. 2 is found 100' east of No. 4, and is about 6' in width; vein No. 1, which also averages about 6', lies 50' east of No. 2.

The first work done on vein No. 2 was the driving of an adit a distance of 300' on the vein. This tunnel was 950' above the bed of Cañon Creek and the ore developed was stoped. At different periods two other adits were driven on the No. 3 vein, one 200' in length, 300' above Cañon Creek; and another 200' in length, 550' above the creek level.

The property was being operated in 1914 by the Columbia Cons. Mines Company, and a lower adit had been started at an elevation of

50' above the creek and was in a distance of 20'; the quartz in the face was said to average $5 to $6 per ton in free gold and 1% pyrite. New buildings had been erected, 2000 inches of water had been secured from Cañon Creek under a 60' head by means of a 1000' flume and a Giant 100-horsepower compressor driven by a Pelton wheel had been installed. The ore will be crushed in a "Bartlett" crushing and grinding machine which has a guaranteed capacity of 250 tons per 24 hours through 20-mesh screen. The nearest producing mine is the Gaston mine which is situated two miles north of the Columbia.

In October, 1915, the old 20-stamp mill of the German Mine was purchased and the machinery moved to the Columbia claim. An aerial tramway has been built from the main tunnel to the millsite. Mill crushing 50 tons per day in December. Compressed air drills are used in drifts.

This property was sold in April, 1916, to the Columbia Cons. Mines Company, for a reputed price of $25,000, the company also taking over the German and Ocean Star properties. The three properties will be worked from the Columbia.

The property was closed August 20, 1918, on account of the high price of labor and material and the difficulty in getting supplies.

**Congo Claim.** (See Sultana.) Owner, Sultana Gold Mining Company, Crocker Building, San Francisco.

Location: Grass Valley District, Sec. 10, T. 15 N., R. 8 E., 3½ miles southeast of Grass Valley. Elevation 2600'.

**Conlon Mine.** Owner, Royal Gold Mining Company, 1102 Claus Spreckels Building, San Francisco; M. H. Herman, president; H. G. A. Brunnier, treasurer and manager, Grass Valley.

Location: Grass Valley Mining District, Sec. 1, T. 15 N., R. 8 E., 3 miles southeast of Grass Valley. Elevation 2900'.
Bibliography: Cal. State Min. Bur. Repts. XI, page 283; XII, page 187; XIII, page 239.

The Conlon or Lafayette vein, belonging to the Osborne Hill vein system, outcrops about 1000' west of the Osborne Hill lode and can be traced on the surface for a distance of 6000 feet. The course of this vein is a few degrees west of north and the dip is from 25° to 42° W. The Jefferson, Winding, Lafayette and Comet patented claims, a total area of 90 acres, cover about 4500' of the vein and comprise the Conlon property; 250' west of the Conlon ledge is the Winding vein, which is parallel in strike and dip.

Prior to 1894, the Conlon shaft had been sunk to a depth of only 190' on the vein, which was of an average width of 26 inches. The vein was also opened 1000' south of the Conlon shaft on the Lafayette claim by means of a 480' crosscut adit, from which drifts were driven 500' south and 80' north, at a depth of 130' below the surface. The vein as developed by this work is said to have averaged 1' in width and the ore to have had a mean value of from \$20 to \$50 per ton. The upper Comet crosscut tunnel 1300' south of the Lafayette adit was driven a distance of 600' to the vein, and good ore is said to have been stoped at this point.

In 1900 active development work was started at the Conlon shaft, but the surface plant was destroyed by fire in September, 1902. It was replaced and by 1904 the shaft had reached a depth of 730' on the vein. From this shaft drifts were driven along the foot-wall on the 200', 300', 500', and 700' levels, the 500' level being driven 750' north and 300' south. The Winding ledge was prospected by a crosscut from the 200' level of the Conlon, and about 1000' of exploratory work was done on this vein. Two ore shoots are said to have been developed: one north of the crosscut, having a length of 100', and the other south of the crosscut 250' in length, both opened by raises and drifts between the 300' level and the surface. Some high grade ore is reported to have been taken from the south pay shoot. From the 500' drift on the Conlon vein a crosscut was driven 247 ft. into the hanging wall, and according to surveys this will have to be extended a distance of 25' to intersect the Winding ledge. It is expected that the north ore shoot, on which a winze from the 300' level was sunk to a depth of 40', will be encountered at the point of intersection of the crosscut and the vein. From this point a drift will have to be driven from 200' to 300' south before striking the downward extension of the south ore shoot.

A crosscut adit was started a few years ago to cut the Comet vein at a vertical depth of 380' below the outcrop. The total length of this tunnel when completed will be about 1500 feet. At the present time it has been driven a distance of 600 feet. The elevation of this tunnel is about 60' lower than the 500' level of the Conlon mine, and the extension of a drift on the vein northward 2000' would not only prospect the lode on the Comet and Lafayette claims, but would when completed drain the Conlon working to a depth of 550 feet. It is said that a large amount of water was encountered in the Conlon shaft just before work was discontinued in 1908. The equipment consists of two 75-h.p. boilers, steam-driven hoist and pump, 3-drill air compressor and a new 10-stamp electrically-operated mill, partially dismantled.

**Consolidated Mine** (Delhi). Owner, Consolidated St. Gothard Gold Mining Company, 407 Front street, San Francisco; F. Klopper, president; B. N. Shoecraft, secretary. Leased to Delhi Mines Company, A. A. Codd, president and general manager, box 703, Reno, Nevada.

Location: Columbia Hill District, Sec. 21, T. 18 N., R. 9 E., 4 miles north of North Columbia. Elevation 2100'–3000'.
Bibliography: Cal. State Min. Bur. Repts. VIII, page 444; XI, pages 305–307; XIII, page 241. U. S. Geol. Survey Folio 66, Colfax. Weed, Mines Handbook, 1918.

The Delhi and St. Gothard mines were consolidated about 1898 and the holdings now cover about 160 acres, including the Last Chance, Yuba River, Yuba River Extension, Helvetia and St. Gothard patented claims. The Delhi mine was worked at intervals prior to 1886 by means of two tunnels. The vein outcrops on the south side of the Middle Yuba River cañon and can be traced from the river (elevation 2100') to a point about a mile south (elevation 3200'). From 1886 to 1893 the mine was worked continuously and the pay shoot had been developed by four tunnels; adit No. 4, 83' above the river and 900' below the outcrop with a length of 1700'; tunnel No. 3, 330' above No. 4, was run 1200'; tunnel No. 2, 200' above No. 3, and tunnel No. 1. By 1893 most of the ore had been stoped above adit No. 4 and the sinking of a shaft below this level was attempted, but owing to the excessive amount of water encountered, it was abandoned and the mine closed down. In 1898 the Delhi was consolidated with the St. Gothard mine, No. 4 adit which was driven as a crosscut for the first 1000' and then on the vein for 1000' was reopened and a 20-stamp mill was erected. Seventy-five feet above No. 4 the vein had split on its dip, 1600' from the portal, the 'East' vein dipping 55° east, and the 'West' vein 80° east. A shaft was sunk on the East vein to a depth of 500' and a crosscut was driven 400' into the footwall before it encountered the West vein. From this shaft levels averaging 800' in length have been run on the East vein at 100-foot intervals. There has been one main ore shoot on the Delhi which has been stoped from the bottom of the shaft to the surface—a vertical distance of 1600 feet. This shoot has varied in length from 250' to 500', and from 6" to 16' in width, averaging 5' to 6'.

In December, 1914, another ore shoot was being developed on the West vein by a drift 100' in length driven from the end of the crosscut on the 500' level. The vein at this point varies in width from 18" to 5 feet.

The ore from the Delhi veins carries free gold, pyrite and some arsenopyrite, the sulphides averaging 1½% and $50 per ton in value. The veins occur in the black Calaveras slates which generally form both walls of the vein; in some cases, however, the hanging wall is said to be 'Porphyry.'

Water is used throughout as power and it is obtained from two sources, 1000" under a head of 263' from the Middle Yuba River by six miles of ditch and flume, and 200" under a 900' head by 8 miles of ditch. The property is equipped with the usual buildings, a Norwalk compressor, capacity 1600 cu. ft. free air a minute, a Giant duplex compressor, capacity 900 cu. ft. free air a minute, and a 20-stamp mill. Thirty-five men were employed in 1914.

[Examined by Hamilton Eddie in 1914.

The mine was closed in 1916, reopened and closed again the same year by local interests. It was later leased to Delhi Mines Company, who attempted to reopen it. The flow of water was said to be 350 gallons a minute when work was stopped in 1914, and apparently the increased flow of water during reopening was responsible for failure to continue work.

The reported production from 1890 to 1914, was $1,514,435.

—*C. A. Logan.*]

**Constitution Mine.** (See National.) Owner, National Gold Mining Company; Peter Taulphans, 1401 Jones street, San Francisco.

Location: Nevada City District, Secs. 9 and 10, T. 16 N., R. 9 E., 2½ miles east of Nevada City. Elevation 3000'.
Bibliography: Cal. State Min. Bur. Rept. XIII, page 249 (Jesse Cons.).

**Crown Point Mine** (William Penn). Owner, Empire Mines Company; Geo. W. Starr, Grass Valley.

Location: Grass Valley District, Sec. 26, T. 16 N., R. 8 E.
Bibliography: Cal. State Min. Bur. Rept. XIII, page 238. U. S. Geol. Survey Folio 29, Nevada City Special.

The Crown Point lode, worked at intervals since 1886, was purchased a few years ago by the Empire Company, since which time no work has been done. The vein, which lies in an altered serpentinized diabase porphyrite near its contact with the Calaveras slates, has been worked to a depth of 620' by means of an inclined shaft and 2500' of drifting. The lode strikes N. 40° W., dips 70° to 80° N. and varies in width from a few inches to 4 feet. The vein carries coarse free gold and from 2% to 3% of sulphides.

**Culbertson Mine.** Owner, Dr. A. T. Tickell, Nevada City.

Location: Graniteville District, Sec. 8, T. 18 N., R. 11 E., 2 miles west of Graniteville. Elevation 4100'.
Bibliography: Cal. State Min. Bur. Repts. XII, page 188; XIII, page 240. U. S. Geol. Survey Folio 66, Colfax.

This mine consists of one claim located about 1890 by M. Culbertson. It was afterwards sold to C. D. Eastin, who sank a shaft 160' in depth. The mine was acquired later and worked by Dr. Tickell, who sank the shaft to a depth of 260 feet. The Culbertson location covers 1500' on the northern extension of the strong Republic-National vein which can be traced from Poormans Creek across the divide to the Middle Yuba River, a distance of 2½ miles. The vein varies in width from 1 to 16', averaging about 6 or 8 feet. It occurs in the Blue Cañon (Calaveras) formation, and both walls are slate. The strike is N. 18° W. and the dip is 60° E. An ore shoot 275' in length, said to average from $7 to $8 per ton, has been stoped from the 260' level to the surface. Drifts

were driven 300' north and south on the 160' level, but little development work was done on the 260' level. A crosscut drain tunnel 350'
in length intersects the shaft at a depth of 75 feet.

Water power is obtained under a 600' head. The mine is equipped
with boarding and bunk houses, blacksmith shop, a hoist capable of
sinking to a depth of 1000' and a '3-drill' Giant air compressor. There
is also a 10-stamp Union Iron Works mill on the property. The mine
was last operated in 1906.

**Daisy Hill Mine.** Owner, North Star Mines Company.

Location: Grass Valley District, Sec. 1, T. 15 N., R. 8 E., 2 miles south of
Grass Valley. Elevation 2800'.
Bibliography: Cal. State Min. Bur. Rept. XIII, page 240. U. S. Geol. Survey
Folio 29, Grass Valley Special.

The vein here which is supposed to be the southern extension of the
Rich Hill vein worked by the Empire mine, was discovered and developed in early days by a shaft 300' in depth. Drifts were run from the
shaft for a distance of 300' both north and south, and the ore shoot
which lies north of the shaft, is said to have been stoped to the surface.
The strike of the vein which occurs in the diabase porphyrite is
north and the dip is 30° west. The property has been idle for many
years. It was recently purchased by the North Star Mines Company.

**Dakota Claim.** Owner, North Star Mines Company.

Location: Grass Valley District, Sec. 34, T. 16 N., R. 8 E., 1 mile south of
Grass Valley. Elevation 2500'.

This fractional claim, containing 6.04 acres, adjoins the W. Y. O. D.
claim on the south. The vein which strikes north and dips to the west
has been developed by a 200' tunnel, but no work has been done for a
number of years. The price paid for this property by the North Star
Mines Company is currently reported to have been about $15,000.

**Deadwood Mine.** Owner, Deadwood Gold Mining Company; J. M.
Hadley, Nevada City.

Location: Nevada City District, Sec. 9, T. 16 N., R. 9 E., 2 miles east of
Nevada City. Elevation 2700'.
Bibliography: Cal. State Min. Bur. Rept. XIII, page 240. U. S. Geol. Survey
Folio 29, Banner Hill Special Map.

The Deadwood, Railroad, and the Oriental Placer mine, including
95 acres of patented ground, comprise the holdings of this company.
From 1856 to 1886 the mine produced $300,000, but very little work
has been done in the last 20 years. The vein outcrops on the north side
of Deer Creek and strikes N. 18° E. and dips 25° W. Both walls are
granodiorite. It varies in width from a few inches to 18 inches, but
carries a high percentage of pyrite, galena, sphalerite and arsenopyrite.
The ore is said to have been high grade, sometimes reaching a value of
$100 to $200 per ton. The vein has been developed by a 50' incline

shaft, from which a number of drifts have been run south a distance of from 300' to 400'.

**Delhi.**  (See Consolidated.)

**Diamond Claim.**  (See Bullion.)   Owner, Bullion Consolidated Gold Mining Company, Alaska Commercial Building, San Francisco.

> Location: Grass Valley District, Sec. 2, T. 15 N., R. 8 E., 2 miles south of Grass Valley.  Elevation 2500'.
> Bibliography: Cal. State Min. Bur. Repts. XII, page 188; XIII, page 241.  U. S. Geol. Survey Folio 29, Grass Valley Special.

**Diamond Creek Mine.**  (See Eagle Bird.)   Owner, Eagle Bird Quartz Mining Company, 309 Lankershim Building, Los Angeles; Wm. M. Wilson, agent, Washington, Nevada County, California; H. S. Mourning, president and manager; L. S. Klinker, secretary.

> Location: Washington District, Secs. 12 and 13, T. 17 N., R. 11 E., 6 miles east of Washington.  Elevation 3500'.
> Bibliography: Cal. State Min. Bur. Repts. VI, pt. II, page 44; X, page 389; XII, page 188; XIII, page 241.  U. S. Geol. Survey Folio 66, Colfax.

**Dinero Mining Company.**  (See also Yuba Consolidated Gold Mining Company.)

This company, incorporated with a capitalization of $2,000,000 by A. T. Hathaway, W. J. Connors, S. Montague and others, of San Francisco, purchased the Washington mine in the summer of 1915. Unwatering of the 640' Yuba shaft was commenced and sinking to an additional depth of 1000' was planned.  Over 16 distinct veins have been encountered, but the main Yuba ledge will be given principal attention.  The property is equipped with hoisting and pumping units and two mills, having a combined capacity of 30 stamps.  (Min. and Eng. World 10/23/15.)

The Yuba group has also been purchased by this company.  Pumps have been installed and unwatering of the 960' shaft will soon be in full operation.  W. J. Connors, Superintendent.  (Min. and Eng. World, Dec. 18, 1915, and Eng. and Min. Jour., Nov. 13, 1915.)

Both properties were idle in September, 1917.

**Dromedary** (Berriman and Rock Roche).  (See Golden Center.)

**Dublin Bay.**  (See Erie.)

**Dutretre Group.**  Owner, Engene Dutretre, Golconda, Nevada.

> Location: Washington District, Secs. 12 and 13, T. 17 N., R. 11 E., 7 miles east of Washington, Nevada County.  Elevation 4000'.
> Bibliography: U. S. Geol. Survey Folio 66, Colfax.

This property consists of four patented claims, the Yuba Cañon, Yuba Cañon Extension, Sunlight and Sunlight Extension, which lie 2000' east of the Eagle Bird mine.  The claims cross the South Yuba River, the elevations varying from 3800' to 5000', and cover about 6000' along a vein which dips 70° E., and is parallel to the Eagle Bird lode. The wall rock of the vein is granite  No work has been done on this property for a number of years.

**Eagle Bird Mine.** Owner, Eagle Bird Quartz Mining Company; W. M. Wilson, agent, Washington, Nevada County, California. Bonded 1914 to California Eagle Bird Mining Company, 309 Lankershim Building, Los Angeles; H. S. Mourning, president and manager; L. W. Klinker, secretary.

Location: Washington District, Secs. 12 and 13, T. 17 N., R. 11 E., 6 miles east of Washington. Elevation 3500'.
Bibliography: Cal. State Min. Bur. Repts. VI, pt. II, page 51; X, page 389; XII, page 188; XIII, page 241. U. S. Geol. Survey Folio 66, Colfax.

The Eagle Bird vein belongs to the Ethel, Baltic and Birchville vein system, which can be traced at intervals from the Eagle Bird northward to Graniteville, a distance of six miles. The holdings consist of the Eagle Bird, Eagle Bird Ext., and Live Oak patented claims, and the Henry Clay and Etna locations, 120 acres of mineral ground, together with 160 acres of timber. The claims cover a length along the lode of 4500' south from the South Yuba river, elevation 3500' to 5000' at the top of the ridge.

The quartz-filled fissure occurs in an area of granite and the walls, which are from 18" to 30' apart, are well defined. The vein has an average width of 8', strikes N. 70° W. and dips 70° to 80° E. The length of the pay shoot was 500' and it was worked from the incline shaft, 850' on the vein, by levels run 200', 300', 400' and 500', below the surface. These levels were driven south a maximum distance of 800', and north 300 feet. A tunnel has also been driven from a point 50' south of the shaft and 50' above it. This adit is 950' in length on the vein and, owing to the steep rise of the cañon's south wall, it has attained a depth of 700' below the outcrop. Very little stoping has been done from the tunnel level. The mine has been worked at intervals since early days. In 1886 the shaft had reached a depth of 400' and in 1890 a depth of 500 feet. After a period of inaction the mine was reopened in 1898, but after a number of years of successful operation, extended litigation caused the mine to be worked only at intervals until 1909. In that year the case was finally decided in favor of the Eagle Bird Company, but a few days after the decision was rendered the entire surface equipment, including the 30-stamp mill, was destroyed by fire. No work was done until the summer of 1914 when the California Eagle Bird Company was engaged in installing new machinery preparatory to unwatering the shaft. The Company owns a water right of 1000' on the South Yuba, and water is available under 180' head.

A 10-stamp mill was being installed in September, 1915, under the supervision of C. J. Klinker. A flume and ditch system to furnish water to the mine for power and mill purposes had also been started.

(The property was idle in September, 1918.)

Photo No. 21. Residence of W. B. Bourn at the Empire Mine near Grass Valley.

**Eclipse Claim.** Owner, Mrs. F. Frank, Grass Valley.

Location: Grass Valley District, Secs. 2 and 11, T. 15 N., R. 8 E., 3 miles south
of Grass Valley. Elevation 2200'.
Bibliography: U. S. Geol. Survey Folio 29, Grass Valley Special.

This claim adjoins the famous Allison Ranch mine on the south, but
very little work has been done on the small vein, which strikes north.

**Electric.** (See Sultana.)

**Emmet.** (See North Star.)

**Empire Mine.** Owner, Empire Mines and Investment Company,
375 Sutter street, San Francisco; W. B. Bourn, president; J. Walter
Ward, secretary; Geo. W. Starr, managing director.

Location: Grass Valley District, Secs. 26, 27, 34, and 35, T. 16 N., R. 9 E.,
1¼ miles southeast of Grass Valley. Elevation shaft collar 2692.97 feet.
Bibliography: Cal. State Min. Bur. Repts. VIII, page 426; X, page 371; XI, page
272; XII, page 189; XIII, page 242. U. S. Geol. Survey Prof. Paper 73.
U. S. Geol. Survey 17th Rept., pt. II, Quartz Veins of Nevada City and
Grass Valley District (Lindgren). U. S. Geol. Survey Folio 29, Grass Valley
Special. Geo. Starr, in Mining and Scientific Press, Aug. 4 et seq., 1900.

The Empire mine is one of the most famous and productive properties
in the wonderful Grass Valley district. Early in 1916, the total holdings were about 430 acres. The history of this mine is typical of
California mines in general, with the exception that it has the distinc-
tion of being one of the few which has never been closed since its
discovery in 1850. The Empire has, however, had its lean and productive periods, but there has never been an actual and complete cessation
of work during 65 years. Since 1887, the Empire mine has had the
benefit of the excellent management of Geo. W. Starr, to whose foresight and business ability the Empire and consolidated mines owe their
era of prolonged prosperity which will undoubtedly continue for many
years to come. Owing to the lesson that the history of this mine
teaches regarding the opportunities offered by abandoned mines in
California, an account will be given in detail.

Following the discovery of the wonderfully rich quartz on Gold Hill
by McKnight in June, 1850, intense activity was inaugurated in the
search for quartz mines and (in October, 1850) in rapid succession
ledges were discovered in Massachusetts Hill, Ophir and Rich Hill.
The first 'homemade' mills which were erected in this district, proved
disastrous failures and for the next few years the quartz mining
industry in Grass Valley was at a very low ebb. The early claims as
located were only 30 by 40 feet and the Empire property as it stands
today consists of hundreds of claims which have been gradually consolidated around the original Ophir claim located in 1850 by Geo.
Roberts. The Ophir, together with adjoining claims, was sold in 1851
to Woodbury & Parks, owners of the Empire custom mill in Boston
Ravine. In 1854 the Empire Mining Company was incorporated; after
purchasing other claims the mine was systematically developed. Owing

Photo No. 21 A. Surface plant of the Empire Mine. 1 and 2, in the distance, the Central Shaft and Massachusetts Hill Shaft of the North Star Mines. 3. Empire Cyanide Plant. 4. Empire Mill. 5. Shops. 6. Headframe.

to the exceptionally rich ore and better management the Empire did
not suffer from the general depression experienced by quartz mining
and the production of bullion as given by the books of the company
are as follows:

| | |
|---|---:|
| October, 1850, to May, 1854 | Unknown |
| May 1 to December 31, 1854 | $152,887 91 |
| January, 1855, to December, 1855 | 192,231 67 |
| January, 1856, to October 31, 1856 | 73,654 74 |
| April 1, 1857, to December 31, 1857 | 72,654 74 |
| January, 1858, to December, 1858 | 112,408 30 |
| January, 1859, to December, 1859 | 121,443 44 |
| January, 1860, to December, 1860 | 102,759 18 |
| January, 1861, to December, 1861 | 72,413 04 |
| January, 1862, to December, 1862 | 96,130 31 |
| January, 1863, to December, 1863 | 59,374 43 |
| | $1,056,234 40 |

The above amount was the yield in free gold from approximately
28,100 tons of ore, giving an average value of $37.59 per ton, not
including the value of the 'sulphurets'.

Photo No. 22.  Headframe and ore bins at the Empire Mine.  Photo by C. A. Waring.

In 1865–1866 the Empire Company, pursuing a policy of expansion,
installed, at a cost of $200,000, a new plant which included a 30-stamp
mill. At this time the quartz mining industry in the Grass Valley
district had entirely recovered from its depression and in 1867 there
were 30 quartz mills in operation and 1600 men employed in the industry, but in the succeeding years, owing to poor management, the lack
of development work and general 'dry rot' this period of prosperity

was again followed by its corresponding period of depression and by 1880 the Empire and the famous Idaho were the only mines operating.

The Empire mill was destroyed by fire in 1870 and was replaced by a 20-stamp mill. By 1878 all the pay ore had been extracted and the 1200' incline on the Ophir vein was allowed to fill with water; work, however, was then undertaken on the Rich Hill vein. The production of bullion from 1865 to October, 1878 was as follows:

| | |
|---|---|
| April, 1865, to December 31, 1865 | $71,780 70 |
| January 1, 1861, to July 31, 1866 | 71,111 31 |
| New 30-stamp mill started. | |
| August 1, 1866, to December 31, 1866 | 85,261 64 |
| January 1, 1867, to June, 1867 | 58,927 80 |
| October, 1867, to October, 1868 | 254,000 00 |
| October, 1868, to October, 1869 | 175,000 00 |
| October, 1869, to October, 1870 | 123,000 00 |
| 30-stamp mill destroyed by fire. | |
| October, 1870, to October, 1871 | 64,000 00 |
| October, 1871, to October, 1872 | 110,000 00 |
| October, 1872, to October, 1873 | 168,000 00 |
| October, 1873, to October, 1874 | 119,000 00 |
| October, 1874, to October, 1875 | 201,000 00 |
| October, 1875, to October, 1876 | 120,000 00 |
| October, 1876, to October, 1877 | 171,000 00 |
| October, 1877, to October, 1878 | 117,000 00 |
| | |
| Production from April, 1865, to October, 1878 | $1,911,081 45 |
| Production from May, 1854, to December, 1863 | 1,056,234 40 |
| | |
| Total production, 1854 to 1878 | $2,967,315 85 |

This amount represents the free gold only, and does not include the sulphides which average 2% to 3%, and vary in value from $60 to $200 per ton.

In 1878 the Empire mine, which at that time was controlled by the W. B. Bourn estate, was examined by three well-known mining engineers who reported that the property was too deep for profitable working and advised that it be closed. At this critical period in the history of the mine, and in fact of the whole district, W. B. Bourn, Jr., after a thorough examination of the mine decided that the property warranted further development. He, therefore, formed the Original Empire Company; development work was commenced, and after energetic and systemic exploration carried forward under the most adverse conditions, the property in 1883 again entered on an era of prosperity. Prior to this time steam had been used as power, but in pursuance of its far-sighted policy of expansion, the company installed water power at an expense of $100,000, reconstructed and improved the surface plant, and in 1886 increased the number of stamps from 20 to 40. In 1887 Geo. Starr, the present managing director, after serving an apprenticeship of six years, became superintendent, but in 1893 he resigned to accept the management of a South African mine. During this time a split had been encountered in the Ophir vein below the 1700' level, and exploratory work was carried on at a loss until 1896 when

Mr. Bourn repurchased the controlling interest, which he sold in 1888. In 1898 Mr. Starr again became manager of the property, and a great part of the prosperity that the Empire mine has experienced from 1898 to the present time, is due to his business-like and efficient management. Before accepting the management Mr. Starr made a complete examination and report on the property, the essence of which is in his opinion applicable to gold mining generally in California today, as it was to the Empire mine in 1898. The report stated that the lack of sufficient development work was the chief cause of all underground trouble and that surface conditions were deplorable, the mine having outgrown the plant. Further, he stipulated as a condition of acceptance of the management that $200,000 be furnished him to properly equip and open the property. The directors at first refused to accept the plan outlined, as they did not consider the amount of ore in sight sufficient to justify such expenditure. Mr. Starr replied that the property had a record of 47 years continuous work and from a very small area of its total holdings, over $7,000,000 had been produced; that the chief cause of failure was lack of systematic development and a plant unsuited for economical working under existing and future conditions; that in his opinion the showing in the mine justified such an expenditure. Mr. Starr's far-sighted policy was at last accepted and the expenditure of $200,000 and a monthly development of 600′ was authorized in May, 1898.

The new plant was completed in 1899 and underground development was rushed. The extension of the 2100′ level proved the length of the ore-body on this level to 1840 feet. Owing to the discovery of rich ore, there was no occasion to use any part of the $200,000 appropriated and the mine produced gold, not only to pay for all improvements, development and operating expenses, but to pay dividends besides. Since 1900 the equipment has been added to from time to time until today the Empire plant, clubhouse and grounds are one of the show places of California. The surrounding property has gradually been acquired until the company now owns over 600 acres of mineral ground, including the Pennsylvania-W. Y. O. D. holdings. This mine had been operated at a loss for over ten years, when Mr. Starr again showed his faith in the Grass Valley district by taking over the property against the judgment of other mining engineers who had thoroughly examined the mine. The output of the Pennsylvania since its acquisition in 1912 and recent (1915) development at a depth of 4600′, have made the Pennsylvania, in the light of future possibilities, a rival of the Empire and North Star mines, and justify Mr. Starr's opinion that the Grass Valley district is one of the greatest gold mining camps in the United States, if not in the world. It is a pleasure to record that the faith, foresight and energy shown by Mr. Bourn and Mr. Starr in developing the Grass Valley district at a time when it was practically abandoned, have been amply rewarded by the dividends paid by the rejuvenated properties.

In 1900 the main Empire shaft was 3080', deep while in 1914 the shaft had reached a depth of 4760', and upon the completion of the new hoist sinking will be continued to a depth of 7500 feet. In the old workings levels were run every hundred feet, but all the lower levels are driven 400 feet apart. In 1914 most of the development work and stoping was being done on the 3000', 3400', 3700', 4200' and 4600' levels from 400' to 3000' north of the shaft. The territory lying south of the shaft is practically unexplored in the deeper levels. A total of 30 miles of drifts and crosscuts have so far been driven to prospect the Empire property, and 26½% of the whole area developed has been profitably stoped. In the lower levels the shaft was sunk in the hanging wall, and crosscuts have to be driven northeast to the Ophir vein. On the 4200' level the vein lies about 4000' from the shaft. Crosscuts have been driven from the 3400' and 4600' levels of the Empire mine to prospect the Pennsylvania lode at an approximate depth respectively of 800' and 2000' below the bottom of the Pennsylvania shaft (2600').

The Empire vein outcrops in the diabase-porphyry and continues in this formation to a depth of 1700', where it enters the granodiorite which forms both walls of the vein to the greatest depth yet attained. The average dip of the Empire ledge from the surface to the 3000' level is 30°, but below 3400 feet the vein steepens and from the 2800' to the 4200' level the average dip is 55°. The walls of the fissure are from 3 to 4 feet apart, but the auriferous quartz will, as a rule, average from 18 inches to 2 feet; in the lower levels, however, a drift has been driven a distance of 200 feet on an ore-body which has an average width of 8' of heavily mineralized quartz. The ore in the Empire mine is characteristic of the Grass Valley district, consisting of quartz carrying fine and coarse gold with 2 to 3% of sulphides which are for the greater part auriferous pyrite with a small amount of finely disseminated galena, usually indicative of high grade ore. The pay-shoots are irregularly distributed through the ore-bearing zone. The greatest 'barren zone' in the mine was found between the 1300' and 2100' levels in which 800' practically no ore was encountered. On account of the irregular distribution of the ore shoots and the barren zones encountered, it is necessary to systematically develop the vein far in advance of the stoping requirements. This policy has resulted in the continued prosperity of the Empire mine, and had it been applied to other mines of the Grass Valley district and throughout the State, many which are today lying idle would be enjoying like prosperity. In 1905, while surveyor at the North Star mines, the writer made sections of the Grass Valley district showing the probable intersection of the Empire-Pennsylvania and the North Star vein systems at a depth of from 6500 to 7500 feet. It is interesting to note that recent developments in the lower workings of the North Star and Pennsylvania mines (4600') have proven that these vein systems do intersect and in the near future further development will in all probability prove that the Pennsylvania-Empire system of fissures are the main fissures of the district, and that they will continue to be ore-bearing to the deepest levels possible to work with present-day methods.

All data regarding production costs and detailed development were refused, but it was currently reported in 1914 that the combined yield of the Pennsylvania and Empire mines was equal, if not greater, than that of the North Star mine, which was $1,200,000 per year. The 40 stamps of the Empire crushed about 54,000 tons per year and the 20-stamp mill of the Pennsylvania about 26,000 tons, or a total of 80,000 tons per year, making the average yield of the ore about $15 per ton. The working costs in all probability varied from $4.50 to $5 per ton in 1914, and the combined properties were therefore probably yielding a profit of $10 per ton, or approximately $800,000 per year. The production of the Empire property from 1850 to 1915 is estimated by Mr. Starr to have been in the neighborhood of $18,000,000.

The plant and equipment of the Empire company is one of the best and most substantial installations in California. The mine is equipped with a Union Iron Works hoist, driven by two 10-ft. Pelton wheels and capable of hoisting 750 tons per 24 hours from a depth of 5000 feet. In December, 1914, a new Wellman-Seaver-Morgan hoist operated by a 500-horsepower Westinghouse motor was installed and is now in use. This hoist is designed for a maximum depth of 7500', and is equipped with cylindrical grooved drums and safety devices. The skips, which hold four tons of ore each, are 14' x 2' x 3' and weigh 3300 pounds, making a total starting load, including weight of rope, of 23,100 lbs. The wooden headframe, containing rock crushers and ore bins, is being remodeled so that the new hoist may be operated without disturbing the old water-driven hoist which will be left in place to be used in case of emergency. There are two Norwalk compressors, one 30" x 30" x 19", and the other 30" x 26" x 17", and an Ingersoll-Rand electrically-driven, direct-connected compressor, capacity 2500 cu. ft. free air a minute. The machine shop and blacksmith shop which are housed in a galvanized iron building with rock foundation and concrete floors, are completely equipped.

[Adhering to its former policy, the company's local superintendent refused any new information regarding underground developments or cost of mining and milling in 1918. It is generally conceded that the ore being mined today (September, 1918) at the Empire is of higher grade than at the North Star, and it is probable that mining conditions are rather better than at the North Star as regards width of vein, and dip, so that underground work is less expensive. The Empire does not pay a bonus as does the North Star, and does not hand pick the waste after hoisting, being content to store coarse waste underground.

The Empire is also much better situated as regards tailings disposal. A cheap earth and rock fill dam is being raised which will solve the

Photo No. 23. Raising the tailing restraining dam at the Empire Mine.
Photo by C. A. Logan.

problem of keeping tailings out of the streams, while the North Star Mines Company have been put to heavy expense to install a bucket excavator to move and store tailings.

Sinking on the incline was stopped near the 4600' level as a result of the North Star-Empire litigation. The vein being followed is said to have steepened up near that level, so that the Empire shaft was going off into the hanging wall, under land to which the North Star Mines Company had agricultural title, and where the Empire Company had no rights except to follow the vein. It is reported that the work in the Empire below the 4600-ft. level is being carried on through a winze from that level.

The suit for damages to the amount of $15,000 which was brought against the Empire Mines and Investment Company, by the North Star

Mines Company, for alleged invasion of its territory and extraction of
ore to that amount, was settled out of court by mutual agreement, an
arbitrary vertical boundary being established.

The Empire has formerly followed a fixed policy of doing 12,000 ft.
of development work annually. It is not known how the present labor
situation has affected this. The smaller mines around Grass Valley,
and the mineral rights under the town have been practically all bought
by the Empire, North Star, and Golden Center companies. The
Empire Company owned 430 acres early in 1916.

The metallurgical treatment at the Empire is simpler than at the
North Star. The 40-stamp mill has been replaced by one containing
60 stamps weighing 1575 lbs. each. These drop 7 inches 102 times a
minute, crushing to 16-mesh and discharging at a height of 7 inches.
They are capable of crushing 5 tons or more each in 24 hours. Inside
and outside amalgamation are used, the pulp going from the plates to
24 Frue vanners. The whole mill tonnage is treated by cyanidation.
The sand is leached and slime is sent to clarifiers, Dorr thickeners,
treated in Pachuca agitators, and filtered with Oliver filters.—
*C. A. Logan.*]

**Empire West Mines** (consolidation of Omaha, Lone Jack, Wisconsin,
Illinois and Homeward Bound mines). Owners, Empire West Mines,
375 Sutter street, San Francisco; E. L. Eyre, president; J. Walter
Ward, secretary.

> Location: Grass Valley District, Secs. 2 and 3, T. 15 N., R. 8 E., 2 miles south
> of Grass Valley by good roads. Elevation 2300'.
> Bibliography: Cal. State Min. Bur. Repts. VIII, page 433; X, pages 373 and 375;
> XI, pages 273 and 284; XII, page 197; XIII, page 258. U. S. Geol. Survey
> Folio No. 29, Nevada City Special.

All of the several properties which now form the Empire West Mines
were located in the early days of the Grass Valley district. The Lone
Jack was located in 1857, and by 1867 the mine had been worked to a
depth of 600', yielding $500,000. The northern extension of the same
vein was worked in the Omaha, which was located in 1865 and yielded
very rich ore. These claims were consolidated in 1885, and were known
as the Omaha Consolidated. The Homeward Bound mine prior to 1896
was only worked to a depth of 350', and the 165' and 268' levels were
driven a maximum distance of 350' south and 750' northward. This
property was included with the Wisconsin and Illinois in a consolidation, the combined properties being known as the Menlo Mining Company. In 1903 the Empire West Mines Company was incorporated
and all of the properties came under one management.

There are three main veins: the Omaha, Menlo-Wisconsin and
Thomson. Very little exploratory work has been done on the latter.
The Omaha vein which has been developed for a distance of 3100' in the
Omaha, Lone Jack and Homeward Bound claims is a quartz-filled

fissure, striking N., and dipping 33° W. The average width of the vein is about one foot. The lode outcrops in the granodiorite, but a short distance north of the Omaha shaft it enters the diabase, the contact on the tenth level being about 20' south of the shaft. On this level a drift in diabase was driven 1000' north where the vein is said to 'split up.' The Omaha shaft is 1575' deep and from 1899 to 1906, 3975' of development work was done on the following levels: 600', 200 feet north; 1100', 300 feet north and 800 feet south; 1400', 350 feet south of Omaha shaft, reaching a depth of 1600', connecting with those from the Omaha shaft. From the Homeward Bound shaft, which is 1000' deep, 1250 feet of drifting was done during the same period, principally on the 1000' level.

The Menlo or Wisconsin-Illinois vein, which lies from 400' to 600' west of and parallel to the Omaha vein, was developed by means of a crosscut 640' in length from the 1400' level of the Omaha mine. This crosscut was also continued for a distance of 90' beyond the Menlo ledge. A drift was driven 500' north of the crosscut, and a raise was put up a distance of 450' on the vein, which averages one foot in width. South of the crosscut the drift was extended a distance of 1500', and a raise of 550' made. From the Wisconsin shaft the 800' drifts were extended a distance of 900' north and 400' south.

From 1899 to 1906 therefore, 10,555' of development work was done on the Omaha and Menlo veins for the most part between the 1000' and 1400' levels over a length of 3000' along the lodes and this work is said to have blocked out 140,000 tons of 'low grade ore'.

The ore is quartz, carrying free gold, and from 3 to 4% of sulphides. From 1890 to 1899, the Omaha Consolidated mine produced 54,966 tons which yielded $883,970, an average of $6.17 per ton; during this period, however, the work was confined almost entirely to development and the ore milled was from tributers, development work and test runs. From 1892 to 1900, inclusive, the Menlo Mining Company produced 10,481 tons yielding $124,496, or an average of $11.87 per ton. It is estimated that the combined properties have produced to date about $3,500,000.

In regard to the future possibilities of the consolidated property, the following is quoted from a report on the mine made in 1906 by Mr. Geo. Starr, manager of the property from 1899 to 1906:

"The results of the 10,555' of development work prove that we have been working in a poor zone of ground and are now in the position of the Empire mine, where that company for eight or nine years labored through a poor zone between the 1500' and 2100' levels. I may also refer to the position of the North Star mines that had a similar zone of poor ground and there are a number of other properties in this district that have had the same experience. I firmly believe that the mines of the Empire West Group are characteristic of the mines in the Grass Valley district, and will in depth prove valuable properties.''

**English Mountain Mine.** Owner, Jas. Holdsworth, Graniteville. Bonded to H. Schroeder, Nevada City.

> Location: English Mountain District, Sec. 6, T. 18 N., R. 13 E., 13 miles east
> of Graniteville. Elevation 6200'–6500'.
> Bibliography: Cal. State Min. Bur. Rept. XIII, page 242.

This mine was discovered in 1868 by Holdsworth, but was worked only superficially until 1894 when it was operated by the English Mountain Gold Mining Company. Three years later the 20-stamp mill was destroyed by fire, and the mine was practically idle until 1908 when the present bondee began exploratory work which has been continued during the summer season of each year. There are five claims in the group of which the Marguerite and Helen D. are patented. The property has an area of 105 acres and covers a length of 4500' along the lode. The country rock is a diabase porphyrite and there are two lodes known as the Big vein and the 'Gold' vein, the latter intersecting the former. The 'Gold' vein, which is a fissure vein in the diabase, strikes northwest and dips 80° southwest. It can be traced on the surface for a distance of 3000', and is said to average from 4' to 5' in width. The tunnels on this vein are driven a distance of 300' each. No. 2 tunnel, 100 feet below No. 1, developed an ore shoot 150' in length and this ore, which in the oxidized zone yielded from $12 to $32 per ton, was stoped to the surface a distance of 150 feet. An adit about 400' below No. 2 tunnel has been driven a distance of 1200', but so far has failed to reach the ore shoot worked on the upper levels on account of the rake of the shoot. There is also a zone of oxidized material called the 'Gossan' vein, which can be traced for a distance of 4500'. A tunnel has been driven a distance of 300' in this at a depth of 250' below the surface, and another tunnel 75 feet lower has been extended 400' in the oxidized zone.

Crosscuts from the tunnels have shown the walls of the ore-body to be from 25' to 50' apart. The gold content is said to range from 80¢ to $37 per ton. The mine has yielded over $100,000, and although it is the only mine in the vicinity which has produced any considerable amount of gold, from reliable information it appears that the district warrants further exploitation and development.

(Owing to the shortness of the season it was found impossible to visit this mine and district, and the information given was furnished by Mr. Henry Schroeder.)

**Erie Mine.** Owners, California Erie Company, holding company; Erie Exploration Company, operating company; E. L. S. Wrampelmeier, president and manager; L. P. Larue, secretary, Grass Valley.

Location: Gaston District, Sec. 21, T. 18 N., R. 11 E., 1 mile north of Gaston, 3 miles south of Graniteville. Elevation 5000'.
Bibliography: U. S. Geol. Survey Folio 66, Colfax.

This property, discovered and prospected in early days, was reopened in 1890 after a period of idleness of 30 years, and in 1900 a new 20-stamp mill was erected. The mine was closed in 1903, but in 1909 it was purchased by the present owners and was in active operation for some years. The property, which covers an area of 320 acres and a length of 4500' along the lode, is composed of the following claims: Canton, Cascaret and Cascaret millsite, Dandarine, Dublin Bay, Erie and Erie millsite, Holland, Holland No. 2 and Holland millsite, Irish Boy and Irish Boy millsite, Lane, McCarthy, Neuralgaline, Oliver and Oliver millsite, Wheeling and Wheeling millsite.

There are two veins, the Erie and a parallel vein, called the Dublin Bay, which lies 1500' west of the Erie. Both of the veins occur in the Calaveras slates a mile west of their contact with the Granite. The Erie and Dublin Bay veins belong to the Gaston system which has been so successfully worked at Gaston. The Gaston vein, however, differs from the Erie vein in that it occurs at the contact of the granite and slates. The Erie vein, which can only be traced at intervals on the surface, strikes north and dips 70° E., while the Dublin Bay vein, which has the same strike, dips 45° E., and can be traced for a distance of 5000 feet.

The orebodies in the veins occur in the form of lenses; in the Erie, the main shoot was 400' in length, varying in width from 5' to 30', with an average of 12 feet; the two shoots developed on the Dublin Bay lode are 70' and 100' in length, and have an average width of from 12 to 15 feet. The ore carries less than 1% of sulphides, which are for the most part pyrite, with some galena and chalcopyrite.

The main working adit, which intersected the Dublin Bay vein 400' from the portal, was driven eastward as a crosscut a distance of 2000', where it encountered the Erie fissure. There are four levels above the adit level which is called No. 5, and 14,000' of drifting and exploratory work has been done, 1600' of which has been driven on No. 5 level. A new 5000' adit was contemplated in 1914 which will give 800' of backs below No. 5 level. The tonnage stoped is not known, but the mine is reported to have produced $500,000 to 1915 and the ore above the 200' level is said to average $8 per ton.

Water was obtained from the North Bloomfield ditch at a cost of 10¢ per miner's inch per 24 hours under 310′ head at the mill. This water was then used again to run the compressors 400 feet below. The mine is equipped with a 12″ x 12″ Norwalk 2-stage compressor of 500 cu. ft. capacity; a Smith Vail 150 cu. ft. compressor, blacksmith shop, tools, drills, cars, etc.; a 20-stamp Risdon mill with 1300-lb. stamps, 2 Wilfley and 1 Johnson concentrator, bunkhouse and other buildings.

The total operating cost, including mining, milling and development was given by Manager Wrampelmeier as $5 in 1915, but the cost of stoping and milling was only $1.32 per ton owing to cheap power and timber. Drifting cost $4 per foot, which included labor, supplies and power. The cost of handling freight from Nevada City is $20 per ton, and due to the condition of the roads in winter all hauling has to be done during the summer months.

(The operating company gave up its option without doing any new work and the mine was idle in September, 1918.)

**Ethel Mine.** Owner, Ethel Gold Mining Company, 241 Sansome street, San Francisco.

> Location: Washington Mining District, Secs. 34 and 35, T. 18 N., R. 11 E., 2 miles east of Gaston and 7 miles northeast of Washington.
> Bibliography: Cal. State Min. Bur. Repts. XII, page 189 ; XIII, page 242.

This mine was idle in 1914 and inquiry at the San Francisco office failed to elicit any information regarding the property.

**Eureka Consolidated.** (See under Copper.)

**Eureka Consolidated Mine** (Roannaise). Owner, Gold Point Consolidated Mines, Inc.,* 1007 Crocker Building, San Francisco.

> Location: Grass Valley District, east of Grass Valley. Elevation 2500′.
> Bibliography: Ross Brown and Raymond Repts. from 1867–1877.

The Eureka mine was located in 1851, and the outcrop of low-grade gold quartz was worked at intervals until 1857. In that year the mine was purchased by Fricot and others, but the ore taken from the surface yielded only $4 per ton, and the mine was worked at a loss until 1863. In that year a vertical shaft was sunk to a depth of 100′, at which point ore averaging $28 per ton was encountered. Between the 100′ and 200′ levels the ore averaged $37, and from there on to the 600′ level it increased in value to $50 per ton over an average width of 3 feet. The mine was sold to the Eureka Company October 1, 1865, for $400,000. The production prior to this time is unknown, but for ten years, from

---

*Since the above was written this property has passed into the possession of the Idaho-Maryland Mines Company, Hobart Bldg., San Francisco.

October 1, 1865, to October 1, 1875, the production and dividends paid were as follows:

| | Tons milled | Value | Dividends |
|---|---|---|---|
| Oct. 1, 1865–Sept. 30, 1866 | 11,375 | $531,431 | $380,000 |
| 1866–1867 | 12,000 | 573,136 | |
| 1867–1868 | 15,944 | 474,184 | 290,000 |
| 1868–1869 | 21,526 | 574,964 | 264,000 |
| 1869–1870 | 20,562 | 661,898 | 400,000 |
| 1870–1871 | 18,560 | 563,055 | 360,000 |
| 1871–1872 | 9,590 | 226,810 | 20,000 |
| 1872–1873 | 7,852 | 486,854 | 260,000 |
| 1873–1874 | 8,130 | 100,021 | 80,000 |
| 1874–1875 | | | |
| Totals | | $4,390,619 | $2,054,000 |

The Eureka vein was marked by a bold outcrop of white quartz which appeared in the serpentine, but as the vein was traced eastward toward the Idaho mine the outcrop became less conspicuous. From the Idaho shaft, eastward it can not be traced on the surface, although one of the largest and richest ore shoots in the Grass Valley district was found extending 4000' east of the Idaho shaft and only a few hundred feet below the surface. In the Eureka ground the vein was developed by an incline shaft, which followed the vein to a depth of 1200 feet. The pay shoot was encountered at a depth of 100', and the shaft continued in rich ore to below the 600' level. From this point to the lowest levels the vein maintained an average width of from 2 to 3', but the tenor of the ore is said to have averaged only $7 or $8 per ton as compared with a value of from $30 to $60 per ton in the main shoot. The Eureka-Idaho-Maryland ore shoot is, according to Lindgren, one of the most remarkable known in vein geology. Compared with the ore shoot of the other Grass Valley mines, and in fact other California mines, it is unique in its remarkable continuity. In the Eureka ground the pay shoot was nearly horizontal beginning at a depth of 100', and ending at a depth of from 600 to 700 feet. This shoot was worked in the Eureka mine for a length of approximately 1600' and was then followed eastward by the Idaho-Maryland workings for a distance of 4000 feet. At a distance of 1000' west of the Eureka shaft the vein was lost and considerable exploratory work, including crosscutting, done in the Roannaise and Morehouse claims, failed to find the westward continuation of the shoot.

This fact, in conjunction with the poor showing of the mine below the 600' level, caused a suspension of work in 1877; all the equipment was removed and no attempt has since been made to explore the property. The total production of the Eureka mine is estimated to have been $5,700,000.

For further details regarding this mine, see Idaho-Maryland mine.

**Excelsior Mine.** Owner, Excelsior Consolidated Mining Company, Worcester, Massachusetts; Henry J. Gray, superintendent.

Location: Meadow Lake Mining District, Sec. 27, T. 18 N., R. 13 E., ½ mile
   west of Meadow Lake and 38 miles northeast, by wagon road and trail,
   from Emigrant Gap. (S. P. R. R.) Elevation 7100'.
Bibliography: U. S. Geol. Survey Folios 39 and 66, Truckee and Colfax.

The Excelsior mine consists of four patented claims, the Excelsior No. 1, Excelsior No. 2, Union No. 1 and Union No. 2. This property is situated in one of the deepest snow regions of the west, and can only be operated economically in the summer. Supplies must be hauled in during the summer from Emigrant Gap. The first discovery of gold in this district was made on this property in 1863.

The vein, of great prominence and width of outcrop, exceeding 45' in places, crosses the property in a northwest and southeast direction. The ore consists of pyrite, arsenopyrite and zincblende, carrying a moderate amount of gold. Free gold occurs only in the decomposed sulphide material.

A new boarding house was erected in the summer of 1914 and by the next summer enough ore was blocked out to warrant the erection of a mill. New equipment was added and a pipe line constructed. By December, 1915, the second unit of three stamps had been added to the mill and a third was in course of construction. A new office building and five residences have been built.

(In 1918 the property was idle. There is a 20-stamp mill, but no activity has been reported for several years, and only a watchman is employed.)

**Fahey.** (See North Star Mines.)

**Fairview Mine** (Calumet and California Mining Company). Owner, Sierra Asbestos Company, Oakland, California.

Location: North Bloomfield Mining District, Sec. 2, T. 17 N., R. 10 E., 2½ miles
   west of Washington, thence 21 miles by good stage road southwesterly to
   Nevada City. Elevation 4325'.
Bibliography: Cal. State Min. Bur. Bull. No. 50, page 202. U. S. Geol. Survey
   Folio 66, Colfax.

The Fairview mine contains 8 claims, having a total area of 160 acres. An open cut on the hill was worked for gold some years ago.

The vein is developed by a 375' adit driven on the vein, giving 170' of backs. Vein matter contains chalcopyrite, red oxide of copper, iron pyrite, gold, quartz and iron oxide. It is 12' wide in places, with a dip of 65° W. and a strike of N. 17° W. The hanging wall is schist, the foot-wall serpentine. Reported values are: Copper 1.1%, gold $4 and silver 16%.

Equipment consists of a car, track, and blacksmith shop.

(This company was reorganized in 1914 as the Calumet and California Mining Company. The property and mill have since been sold to Sierra Asbestos Company.) (See under Asbestos.)

**Federal Loan Mine.** Owners, Joseph Weisbein Estate and Jacob Weisbein, Nevada City.

Location: Nevada City District, Sec. 10, T. 16 N., R. 9 E., 2½ miles east of Nevada City. Elevation 2800'.
Bibliography: Cal. State Min. Bur. Repts. XI, page 290; XII, page 189; XIII, page 243. U. S. Geol. Survey Folio No. 29, Banner Hill Special.

The Federal Loan mine was worked in early days, but was not opened on a large scale until 1890. In the following eight years the shaft was sunk to a depth of 800 feet. The mine was purchased by the present owners in 1898, and operated by them until 1906, when it was closed and has since remained idle. The vein lies in the siliceous clay slates of the Calaveras formation near their contact with the Nevada City area of granodiorite. The slates have been altered near the contact by metamorphic action of the intruded granodiorite. The vein, which strikes N. 80° E., and dips 45° S., varies in width from a mere seam to a maximum width of 4', probably averaging 1 foot. The filling consists of milling quartz, together with stringers of quartz and altered wall rock. The ore contains free gold and from 4 to 6% of sulphides consisting of pyrite, arsenopyrite and smaller amounts of galena and chalcopyrite. Its average value is stated to have been $15 per ton.

The lode was developed by means of an incline shaft on the vein, which has attained a depth of 1000 feet. Lateral development consists of drifts driven on the vein from 100' to 700' east and west of the shaft on levels 5, 6, 7, 8 and 9, and 75' of drifting on the 1000-foot level. The main ore shoot, while irregular in shape, averaged about 200' in length and one foot in width. Very little ore has been stoped below the 700' level. The mine is credited with a production of $200,000. Water from Deer Creek is used for power. There is no equipment.

**Fillmore.** (See Sultana.)

**Forlorn Hope.** (See Black Bear.)

**Fortuna.** (See Mountaineer.)

**Fountain Head.** (See under Drift.)

**Fruitvale Mine** (Andy Fitz). Owner, Fruitvale Mining Company; care of Andrew Fitzgerald, Moores Flat, Nevada County, California.

Location: Bloomfield District, Sec. ?, T. 18 N., R. 10 E., 4 miles northeast of Bloomfield.
Bibliography: U. S. Geol. Survey Folio No. 66, Colfax.

Work resumed in February, 1914. The tunnel on the vein has reached the 700' point at a vertical depth of 400'. (Idle in September, 1918.)

**Gambrinus.** (See Gaston Gold Mining Company.)

**Gautier Ranch.** (See under Copper.)

**Garage Claim.** Owner, A. B. Snyder, Grass Valley.

Location : Nevada City Mining District, Sec. 27, T. 16 N., R. 8 E., located in town of Grass Valley.
Bibliography: U. S. Geol. Survey Prof. Paper No. 73, W. Lindgren; pages 125–132. U. S. Geol. Survey 17th Ann. Rept., pt. II, pages 1–262, 1896. U. S. Geol. Survey Folio 18, Smartsville; also Folio 29, Nevada City Special.

Ore taken from a 100′ shaft sunk under a garage, has returned $8 per ton, but it is claimed this is not profitable ore. Will sink another shaft 100′ in belief that better ore will be found. Owner is restricted to a depth of 100′, as mineral rights below that depth are owned by Golden Center of Grass Valley Mining Company, who have lately been doing considerable prospecting on the vein at an incline depth of 500 feet, through a crosscut from the Golden Center shaft.

**Gaston Mine** (Gambrinus and California). Owner, Gaston Gold Mining Company, Merchants Exchange Building, San Francisco; E. J. McCutcheon, president, Merchants Exchange Building, San Francisco; Geo. W. Starr, managing director, Grass Valley.

Location : Washington Mining District, Secs. 21, 22, 27, 28, and 33, T. 18 N., R. 11 E. Town of Gaston 6 miles northeast of Washington. Elevation 3500′–5500′.
Bibliography: Cal. State Min. Bur. Repts. VIII, pages 448 and 451; XII, page 189; XIII, page 244. U. S. Geol. Survey Folio 66, Colfax.

The first authentic report on the Gaston mine is that of the State Mining Bureau in the 8th Report (1888), at which time the Gaston Ridge mine, locally known as the California, had been developed to a depth of 147′ by a 600′ tunnel and another tunnel 100′ below was nearly completed. A ten-stamp mill had recovered $45,000 from August 12, 1887, to June 14, 1888. From 1888 to 1898 the Gaston mine was worked intermittently, but by the latter year the vein had been worked to a depth of 275 feet. In 1898 the Gaston Gold Mining Company began active operations. A main adit was started 400′ below the upper workings, and in 1900 at a distance of 2000′ it intersected the Gaston vein. This tunnel was continued 600′ into the hanging-wall where in 1901 the Gambrinus vein was encountered. Later, a winze was sunk on the Gaston vein, and some of the best ore that the mine produced was taken out from below the adit level. In the eight years from 1899 to 1907 the production was approximately $1,000,000 from 174,000 tons of ore stoped from both the Gaston and Gambrinus veins. In 1905 a campaign of energetic development was inaugurated and new equipment at the cost of $50,000 was ordered. A drain tunnel was started 800′ vertically below the main adit, but owing to the San Francisco fire it was found impossible to secure capital to carry out the extensive plans for improvement. All development work was suspended, the ore in sight was stoped, and the mine was closed temporarily in the fall of 1906. In 1908 work was resumed and the drain tunnel was driven a distance of 4500′ to the Gaston veins. A raise was put through connecting the drain tunnel and the bottom of the winze 500′

above. After ten stamps of the 40-stamp mill had been removed to the lower tunnel and run on ore from the lower levels, the mine was closed in the fall of 1913. Work was again resumed in the spring of 1914 and the mine was being successfully operated in 1915 with 25 men employed.

The property owned by the Gaston Gold Mining Company consists of 1200 acres (750 acres patented), including 26 mining claims; of the holdings, 100 acres are heavily timbered.

There are eight separate and distinct veins on the property, but only two, the Gaston and Gambrinus, have been worked.

The Gaston vein varies in width from 2' to 10', dips 40° to 60° E., and strikes N. 20° to 35° W. It occurs at or near the contact of the Calaveras slates and granite, the former forming the foot-wall and the latter the hanging wall. The ore-bodies have been formed by the filling of the deep-seated, persistent fault fissures, and in part by a replacement of the wallrock, by white quartz carrying gold, galena, pyrite and chalcopyrite. South of the main adit a split occurs in the Gaston vein, a 10' foot-wall vein followed the contact and a 4' hanging wall vein swung into the granite. The ore varies from $2 to $50 per ton, with an average value of $10. Both of the veins were worked for a distance of 400' south of the main adit and at that distance they were about 100' apart. The shattered mass of granite lying between the veins has been altered by the vein solutions and this 'blue quartz' is said to vary in value from $1 to $20 per ton with an average value of $4.50. In some places in the mine the altered and silicified granite has been stoped for a width of from 50 to 75 feet.

The Gaston orebodies were worked by means of a winze, to an inclined depth of 500' below the adit level, and produced the major part of the 175,000 tons worked from 1899 to 1907.

The lower or drain tunnel cut the Gaston vein at a depth of 800' vertically below the adit tunnel, but the vein at that point was broken up. Subsequent development work on the lower level, including the raise, put through to the bottom of the winze, was done in the practically barren hanging wall of the main vein. On this account the mine was closed in the fall of 1913. In the spring of 1914 at the instigation of a former foreman, a crosscut, only a few feet in length was driven from the south drift, into the foot-wall, disclosing a vein of quartz over 6' in width. When the mine was visited in August, 1914, a drift had been driven for a distance of 100' on this vein and the quartz from the face showed free gold. A recently erected 10-stamp mill was crushing ore taken out in development work. Those in charge, however, refused to give any information regarding costs of production.

The Gambrinus vein, which averages from 4' to 6' in width, has a general strike of N. 20° E. and a dip of 70° E. The main tunnel encountered the the Gambrinus vein 600' beyond the Gaston vein, but owing to the difference in their strike, at the southern extremity of the workings the veins are only 200 feet apart, and they will probably intersect within a few hundred feet. The ore stoped above the main adit level ran from $10 to $40 per ton. So far the work done on this vein from the Drain tunnel level has failed to develop any ore-bodies of economic importance.

The ore is free milling and carries from 1% to 2% of sulphides which have a value of from $50 to $75 per ton. The average cost of mining and milling up to 1914 was from $2.50 to $3 per ton, owing to cheap power, abundant timber and to the fact that in stoping comparatively little timbering is required.

The property is completely equipped with a plant which cost over $200,000 and which is, for the most part, in good condition. The equipment includes all necessary buildings, a 10-stamp mill and a Hardinge mill with capacity equal to an additional 10 stamps, blacksmith shop and machine shop, hydro-electric power plant (120 K. W. A. C.) auxiliary steam electric plant (75 K. W. A. C.) and an Ingersoll-Rand Duplex Class D. 28¼" x 17¼" x 24" compressor, belt connected to a 6' Pelton water wheel housed in a corrugated iron building, which is located at the mouth of the lower drain tunnel.

Water for power and domestic use is obtained from Bowman Lake, the reservoir of the Northern California Water Company, whose Bloomfield ditch runs through the Gaston property at an elevation of 400' above the mouth of the main adit. The water is first used under this head for running a Rand 16" x 22" Class D compressor, at the mill and is then used again at the drain tunnel level 200' below to operate the electric generator and No. 2 compressor. When the new Ingersoll-Rand compressor was installed at a cost of $20,000 at the drain tunnel level, an independent pipe line was installed from the Bloomfield ditch and the water was applied to the 6' Pelton wheel under a head of 1150 feet. Water is furnished at 10¢ per miner's inch and the cost of power therefore was only about $400 per month when the mine was operating.

The Gaston is the most promising mine in the Washington-Graniteville district and with competent management and systematic development will again become one of the large producers of Nevada County.

[The Gaston mine was closed in the spring of 1918, presumably for the period of the war, on account of high cost of operation. In September, 1918, the superintendent, W. L. Williamson, reported that two men were breaking rock and that 100,000 tons of ore, said to average $4.00 a ton, were then blocked out. The working tunnel had reached a length of 4876 feet. The mine is equipped with a 10-stamp mill, and a Hardinge mill equivalent to 10 stamps more. At the time operations were suspended, the superintendent said the total cost of mining and milling including overhead, had reached $3.92 a ton.— *C. A. Logan.*]

**General Grant Claim.** Owner, General Grant Mining Company; A. F. Perrin, Grass Valley.

Location: Grass Valley District, Sec. 14, T. 15 N., R. 8 E., 3 miles south of Grass Valley. Elevation 2000'.
Bibliography: Cal. State Min. Bur. Register of Mines, Nevada County.

This property consists of one patented claim, the General Grant. No work was being done in 1914. (Idle in September, 1918.)

**German.** (See Bluebell.)

**Giant King Mine.** Owner, Giant King Mining Company, Washington, Nevada County, California; Mr. Metzner, manager.

Location: Washington Mining District, Sec. 13, T. 17 N., R. 10 E., 2 miles south of Washington. Elevation 3100'.
Bibliography: Cal State Min. Bur. Rept. XIII, page 245.

This mine was being operated in 1914, but the manager was not at the mine when the district was visited.

Letters addressed to the company at later dates failed to bring a reply. (Subsequently, the mill and other machinery were sold to the Columbia Consolidated Mines Company, and the property was idle in September, 1918.)

**Gladstone Mine.**

Location: Sec. 2, T. 15 N., and Sec. 34, T. 16 N., R. 8 E., 1 mile south of Grass Valley.
Bibliography: U. S. Geol. Survey Prof. Paper 73, W. Lindgren; pages 125–132. U. S. Geol. Survey 17th Ann. Rept., pt. II, pages 1–262, 1896. U. S. Geol. Survey Folio 18, Smartsville; also Folio 29, Nevada City.

Installation of a new electric hoist designed to sink 2500' and operate at a speed of 1000' per minute with a load exceeding 2 tons, was completed in August, 1914. It is equipped with a 275-horsepower motor. (The property was idle in September, 1918, and no late activity was reported.)

**Glencoe Claim.** Owner, A. Nivens, Nevada City.

Location: Nevada City Mining District, Sec. 18, T. 16 N., R. 9 E., 1 mile south of Nevada City. Elevation 3000'.
Bibliography: Cal. State Min. Bur. Rept. XIII, page 245. U. S. Geol. Survey Folio 29, Nevada City Special.

The Glencoe claim controls 1500' along the eastern end of the Orleans-Gracie-Glencoe vein, which can be traced from the southern extremity of the Champion vein system, S. 70° E., for a distance of 2½ miles. This vein, dipping 75° S., occurs in the Calaveras slates and has only been developed superficially in the Glencoe and adjoining Gracie, Orleans and Fortuna mines to the west. No ore shoots of economic importance have been found by the exploratory work.

On the Glencoe claim the vein has only been opened to a depth of 100′ by a 500′ drain tunnel and a shaft. It is said to have an average width of 4 feet. Adjoining mines are the Gracie, Orleans, Fortuna and Canada Hill, all of which were idle in 1918.

**Glenn.** (See Rocky Glen.)

**Gold Bar.** (See North Star.)

**Gold Canyon Claim.** (See Sierra County—Gold Canyon.) Owner, Mrs. C. Hill, 37 Croxton avenue, Oakland. Bonded to Gold Canyon Mines Company of Nevada City; Edwin T. Blake, president, Oakland Bank of Savings, Oakland, California; Chas. C. Derby, manager, Nevada City.

> Location: Graniteville-Alleghany Mining District, Sec. 11, T. 18 N., R. 10 E., 3 miles south of Alleghany, Sierra County. Elevation 3000′.
> Bibliography: U. S. Geol. Survey Folio 66, Colfax.

**Gold Flat.** (See Pittsburg.) Owner, Pittsburg Gold Flat Mines Company, Nevada City, California.

**Gold Flat.** (See Mohigan.) Owner, Geo. B. Finnigan, agent, Nevada City.

**Gold Hill Mine.** Owner, North Star Mines Company, New York.

> Location: Grass Valley Mining District, Secs. 27 and 34, T. 16 N., R. 8 E. Elevation 2500′.
> Bibliography: Cal. State Min. Bur. Register of Mines and Minerals, Nevada County. U. S. Geol. Survey Folio 29, Nevada City Special. U. S. Geol. Survey 17th Ann. Rept., pt. II.

This mine was reopened by the North Star Company about 1902, but after a number of years of unsuccessful exploratory work it was closed and has since remained idle. For the past history and figures of production of this mine, see under North Star Mines Company.

**Gold King Claim.** Bonded to Henry J. Snyder.

> Location: Grass Valley Mining District.
> Bibliography: U. S. Geol. Survey, W. Lindgren, Prof. Paper 73, pages 125–152. U. S. Geol. Survey 17th Ann. Rept., pt. I, pages 1–262, 1896. U. S. Geol. Survey Folio 18, Smartsville, and 29, Nevada City Special.

This claim was taken under bond by Snyder in the summer of 1915. Good ore is exposed at several points. (Idle in September, 1918.)

**Gold Mound Mine** (Roach and Boyle placer mine). Owner, Gold Mound Mining and Milling Company, 403 Empire State Building, Spokane, Washington. H. Reber, president, Kenosha, Wisconsin; J. F. Bolster, secretary, Spokane; Howard B. Dennis, manager, Grass Valley.

> Location: Rough and Ready Mining District, Sec. 29, T. 16 N., R. 8 E., 2½ miles west of Grass Valley.
> Bibliography: U. S. Geol. Survey Folio 18, Smartsville.

The ground held by this company was originally located and patented as placer claims by Boyle and Roach. In 1903 the Lucky Boy, West Virginia, Pleasant and Taft quartz claims, were located, covering part

Photo No. 24. Outcrop of the Giant King vein near Washington.

of the patented placer ground. The holdings of the company cover about 100 acres in all and 3000' along the course of the vein, which strikes N. 35° E. and dips 57° W. The hanging wall is granodiorite and the foot-wall slate. The vein has an average width of 30 inches, but in some places it reaches a maximum of 8 feet. The Gold Mound vein has been developed to a depth of 270' by an inclined shaft and levels have been run at depths of 100, 150 and 250 feet. On the 100' level drifts were driven on the vein 220' north and 180' south; on the 150' and 250' levels they have been driven only to the south, for distances of 300' and 146' respectively. On the 100' level three short ore shoots were opened; on the 150' level the vein is said to average 30 inches in width from the shaft to the face of the drift; on the 250' level an ore shoot was encountered 20' south of the shaft and the drift is said to be in ore to the face. The ore on the 150' level is reported as averaging $12 per ton, while the ground so far opened on the 250' level will average $6 per ton. The vein has been stoped from the 150' to the 100' level.

Gasoline is used for power and the equipment consists of a 40-horse-power Foose engine, which drives an Ingersoll 2-drill 10" x 10" compressor, furnishing air for the 15-horsepower Fairbanks-Morse hoist, a 6" Cornish pump and the blacksmith shop.

The manager reported early in 1919 that war conditions had prevented any new work for the past four years.

**Gold Point Group.** Owner, Gold Point Consolidated Mines, Inc.,[*] 1007 Crocker Building, San Francisco.

> Location : Grass Valley Mining District, Sec. 25, T. 16 N., R. 8 E., 2 miles east of Grass Valley. Elevation 2500'.
> Bibliography : Cal. State Min. Bur. Rept. XIII, page 245. U. S. Geol. Survey Folio 29, and 17th Ann. Rept., pt. II. The Gold Quartz Veins of the Nevada City and Grass Valley Districts, W. Lindgren.

A consolidation early in 1919 brought the Eureka, Roannaise, Idaho-Maryland, South Idaho, Gold Point, Union Hill and Tracy under this company's ownership. For reports on these, see under the individual names of the several mines.

The Gold Point property consists of three patented claims known as the Gold Point, Centennial and Halphene, which comprise an area of thirty acres in the renowned Eureka-Idaho-Maryland territory. This part of the Grass Valley district has produced more gold than either the North Star or Empire areas, but on account of lack of capital the Gold Point vein has only been superficially developed to a depth of less than 150 feet.

The outcrop is one of the strongest in the district and it can be traced on the surface for a distance of over 2000 feet. The strike of the Gold Point vein is N. 40° W. and the dip is 70° SW. The vein occurs in amphibolite schist which also forms the wall of the Brunswick vein, a

---

[*] Since the above was written this property has passed into the possession of the Idaho-Maryland Mines Company, Hobart Bldg., San Francisco.

parallel vein lying 600' in the foot-wall of the Gold Point. The Union Hill lode which occurs 400' in the hanging wall of the Gold Point and is parallel to it in strike and dip, also lies in this area of amphibolite schist.

The development work on the Gold Point claims consists of a 500' crosscut adit, driven in an easterly direction from the banks of Wolf Creek. At the point where the crosscut adit encountered the vein, a drift was driven northwest on the vein a distance of 600', at which point a connection was made with the inclined shaft from the surface. This shaft and drift have prospected the vein only to a depth of from 100 to 150' below the outcrop. In these workings the vein varies in width from 2' to 6', averaging about 4'. In the foot-wall there is a so-called 'back-vein,' 18" in width, from which specimen ore has been taken. According to Lindgren, this exploratory work opened up a "large body of low grade ore," ranging in value from $2 to $8 per ton. So far, however, no high grade ore-shoot has been encountered. The exploratory work done, however, is insufficient to prove the existence or non-existence of the extremely rich ore-shoots which have made the adjoining mines world-famous. The showing so far made certainly warrants further development, and the likelihood of opening up a pay-shoot on the Gold Point vein is as good as in the case of the Eureka and Idaho mines at a similar stage of their development. (See description of the Eureka and Idaho Mines.)

The mine is without equipment at the present time.

**Gold Tunnel.** (See California Consolidated.)

**Golden Chain.** (See Syndicate Consolidated.)

**Golden Center Group** (Rock Roche, Dromedary and Berriman Consolidated. Owner, Golden Center of Grass Valley Mining Company, 607 J street, Sacramento, care of White, Howard & McCormack. E. R. Abadie, Jr., superintendent, Grass Valley.

Location: Grass Valley Mining District, Sec. 27, T. 16 N., R. 8 E., in townsite of Grass Valley. Elevation 2590'.
Bibliography: Cal. State Min. Bur. Rept. XIII, page 235.

The Golden Center property is a consolidation of Dromedary-Rock Roche and Berriman Consolidated, together with the mineral rights under the townsite of Grass Valley, an area of 150 acres. The Dromedary, Hill, Whiskey, Church Hill, and Rock Roche veins outcrop on the Golden Center holdings. It is anticipated that the northern extensions of the numerous west-dipping veins which outcrop in the territory east of the Golden Center, namely the Kate Hayes, Pennsylvania, W. Y. O. D., Parr, Cassidy and Empire veins will be encountered by the proposed vertical shaft to be sunk on the Golden Center property. (It is claimed that the companies which own these claims to the east have

13—46900

no extralateral rights to these veins under the Golden Center property. According to a map and cross section by F. M. Miller, Grass Valley, these veins will be intersected at the following depths: the Kate Hayes 450′, the Pennsylvania at 780′, the W. Y. O. D. at 990′, the Parr at

Photo No. 25.   Headframe and bins, Golden Center Mine,
Grass Valley.   Photo by C. A. Logan.

1300′, the Cassidy at 1500′, and the Empire veins at 2000′). The continuation, width, and tenor of ore of these veins in the Golden Center property are of course problematical. The Dromedary vein, worked first in the 50's was reopened and worked with considerable success from 1863 to 1874; the ore during this period varying in value from $10 to $60 per ton and the vein varying in width from 1 to 3'. The shaft was sunk to a depth of 300' during this time, drifts were run north and south on the vein and a 5-stamp mill was erected. The Dromedary-Rock Roche mine and the adjoining property were acquired by the present company in 1912 (the property was equipped with small hoist, pump and 10-stamp mill), the old shaft was reopened, and considerable development work was done on the Dromedary veins. In these workings some rich rock was encountered. A crosscut has now been run eastward from the inclined shaft and a vertical raise is being put through to the surface as part of the proposed vertical shaft. When this work has been completed, a new hoist and other equipment will be installed and the sinking of the shaft will be energetically prosecuted.

A new 10-stamp mill was installed and in operation early in 1914, and the inclined shaft had been sunk to a depth of 700'. A crosscut to the Whiskey vein from the 700' level showed 16" of ore. The Garage vein was intersected at about 425' and showed milling gold quartz. The Church Hill vein was being developed from the 250' level. Early in 1915 a 500-gallon pump was installed to cope with the water, which was flowing about 340 gallons per minute. In January, 1916, a 475-gallon per minute pump, driven by a 40-horsepower motor was added to those already lifting 1100 gallons.

During March, 1916, the rich shoot encountered at several points in the upper levels, was encountered at a depth of 850'. It was very strong at this depth, with streaks of high grade showing.

During the year ending March 1, 1916, the company did 2200' of development; crushed 7969 tons of ore, recovered gold worth $112,352 and concentrates valued at $21,638, a total of $133,991, or an average of $16.81 per ton; paid $58,000 for property, etc.; erected a plant costing $35,401; made a profit of $49,843 and paid $21,990 in dividends.

[Since the above report was written the inclined shaft, which dips west at an average angle of 28° following the vein, has been sunk to 1160 feet. The main ore-shoot has been stoped from the 500, 600, 700, 750, 800 and 900-ft. levels north of the shaft. This shoot is crossed by the shaft at 1100-ft. inclined depth. There is said to be good ground unmined on and below the 800-ft. level, north of the shaft, and the main shoot has been only slightly explored below the point where it is crossed by the shaft. Drifts to the north were carried about 500 feet on the 900-ft. level and over 700 feet on the 800-ft. level and the vein has been stoped out as far as drifted on these levels. The 700 north drift has been driven considerably farther and 16" of good ore was being stoped there in September, 1918. Considerable stoping has also been done between the 500-ft. and 700-ft. levels. A crosscut driven northwest from the 500-ft. level encountered the Garage vein, 20″ wide, and this vein has been followed for 500 feet, but has not yet been stoped. South of the shaft some stoping was done in 1915 from the 600-ft. level, and a little in 1917 from the lower levels, but operations in this direction have been limited. Under the last management attention was directed mainly to the one rich ore shoot. Little advance development work was done and there remains a great deal of promising ground as yet unexplored.

In a crosscut from the shaft at the 1100-ft. level in the direction of the east-dipping Peabody vein a heavy flow of water was encountered only a short distance from the shaft. This drove the miners from the lower levels and rose nearly to the 700-ft. level, but in September, 1918, a sinker pump of large capacity was about to be installed and it was confidently expected that complete unwatering was only a matter of reasonable time. Meanwhile development work is going forward on the 500-ft. and 700-ft. levels. The problem of handling water is aggravated in the Grass Valley district by the complexity of the vein systems and it sometimes happens that a company in sinking, taps accumulated supplies or finds itself working in comparatively dry ground where some neighbor has sunk deeper in the immediate vicinity.

The surface plant includes a wooden headframe about 100 feet high, with hoist, ore bin, Gates gyratory and one jaw crusher; two compressors, one Duplex Ingersoll-Rand with capacity of 1200 cu. ft. of free air per minute, driven by a 150-horsepower motor and the smaller a Laidlaw-Dunn-Gordon.

The mill contains 20 stamps, 1500 lbs. each, and 5 Frue vanners. Sulphides average 2% of ore content and $50 a ton value. A force of 24 men, including 4 sets of tributers, were working at time of visit. Past operations have yielded ore which averaged a handsome profit over cost of operation, which was $6 to $7 a ton. To properly develop the mine with a view to the future, the shareholders should expect to forego immediate dividends and be willing to keep exploratory work well ahead of stoping.

The Cabin Flat and Peabody mines have been acquired by this company since the 1915 report was written.—*C. A. Logan.*]

**Golden Gate Group** (Alpha-Kentucky). Owner, Golden Gate Consolidated Mines Company, 404 Mills Building, San Francisco; Chas. C. Haub, president, 45 Powell street, San Francisco.

Location: Grass Valley Mining District, Secs. 25 and 26, T. 16 N., R. 8 E., 1 mile east of Grass Valley. Elevation 2600′.
Bibliography: U. S. Geol. Survey Folio 29.

The Golden Gate property includes the Dana, Christopher Columbus, Golden Gate, Alpha, and Ismert Ranch (agricultural) patented ground,

Photo No. 26.  The Golden Gate Mine, looking north from the Idaho-Maryland.

a total area of 230 acres. The various veins on the claims, together with the Spring Hill vein, form a complex system of interlinked veins, lying about 1420' north of the Eureka-Idaho-Maryland vein. These veins, known as the Golden Gate, Dana-Mobile, Treasury, Alpha, Blight and Wills, occur in the area of serpentine, which forms the foot-wall of the south-dipping Eureka-Idaho-Maryland vein. The dip of the Golden Gate veins is 30° to 34° N., and they thus form the conjugate system to the Idaho-Brunswick south-dipping veins. The general strike of the veins is east and both walls are serpentine. Dikes of diabase have been encountered, and mariposite occurs in the altered serpentine. The veins are irregular in width and the orebodies of quartz are more or less lenticular in shape. The serpentine hanging wall is heavy and difficulty is experienced in keeping the workings open, owing to swelling ground. The ore consists of quartz containing free gold and from 1% to 2% pyrite, the latter worth $150 per ton. The main pay shoot had an average length of 220' and in places the vein reached a width of five feet; the average, however, is about 18 inches.

The Golden Gate veins have been developed by a 1020' inclined shaft, giving a vertical depth of 498' below the outcrop. On the 200' level 840' of drifting was done; on the 250' level, 195'; on the 300' level, 65'; on the 400' level, 1010'; on the 600' level, 75'; on the 850' level, 540'; and on the 1030' level, 590', making a total of 3315' of drifts on the veins. In addition, 485' of crosscuts and 420' of raises and 160' of winzes have been driven. The mine was closed in June, 1914, at which time the ore in sight consisted of about 1000 tons, having an average value of $4 per ton.

The property is completely equipped, including a Laidlaw-Dunn-Gordon compound 600-cubic foot air compressor, double-drum electric hoist, electric turbine and 3 Worthington pumps, 2 small compressed air hoists, 3 electric motors of 100, 75 and 40-horsepower, blacksmith shop, tools, cars, drills, etc. The reduction equipment consists of a 10-stamp mill (1000-lb. stamps), rock-breaker, concentrators and three electric motors of 20, 15 and 5-horsepower, for operating the above. Electric power is obtained from the Pacific Gas and Electric Company, at a cost of 1¼¢ per K. W. H., or a total cost of from $400 to $500 per month for power.

**Good Hope.** (See Scotia.)

**Goodall Group.** Owners, J. M. Thomas et al., Grass Valley, California.

> Location: Grass Valley Mining District, Sec. 1, T. 15 N., R. 8 E., 3½ miles southeast of Grass Valley. Elevation 2700'.
> Bibliography: Cal. State Min. Bur. Rept. XIII, page 246. U. S. Geol. Survey Folio 29, Grass Valley Special.

There are four patented claims in this group known as the Goodall, Governor Perkins, Manhattan and South End, 80 acres in all. No work of importance has been done on the property in the last 20 years. The Goodall vein, which can be traced on the Goodall and Manhattan claims, at intervals, and the South End vein, which is the southern continuation of the Conlin-Lafayette-Comet vein both belong to the Empire-Osborn Hill vein system of west-dipping fissures. There are two parallel veins on the South End claim, which have been developed by two tunnels, one of which is 500' in length and 100' below the outcrop and the other 250' long and 60' below the surface. The vein occurs in diabase-porphyrite and averages 2' in width; the general strike is north and the dip is 60° W. The Goodall vein lies 500' west of the South End vein and has been developed to a depth of 50' by a crosscut adit 500' in length. The veins in this section of the Grass Valley district warrant further prospecting. There is no equipment on the property.

**Governor Perkins.** (See Goodall.)

**Granite Hill.** (See North Star.)

**Gracie Claim.** Owners, Estate of John Arbogast, F. L. Arbogast, Nevada City.

Location : Nevada City Mining District, Sec. 18, T. 16 N., R. 8 E., 1 mile southeast of Nevada City. Elevation 2800'.
Bibliography : U. S. Geol. Survey Folio 29, Banner Hill Special.

The Gracie patented claim covers 1500' along the Gracie-Glencoe vein, which lies in the narrow belt of Calaveras slate between the diabase on the south and the granodiorite on the north. The strike of the vein is N. 80° W. and the dip 85° south. The outcrop of this vein is one of the most continuous in the district and can be traced from the Glencoe, which adjoins the Gracie on the east, for a distance of three miles westward. The vein is said to average 2' in width, and it has been developed by a prospect shaft 100' deep. The nearest mines are the Pittsburg and Canada Hill, which have been idle for a number of years.

**Grant.** (See Canada Hill.)

**Grant Claim.** Owner, J. O. Boyd, San Jose, California.

Location : Grass Valley Mining District, Secs. 25 and 26, T. 16 N., R. 8 E.,
1 mile east of Grass Valley. Elevation 2500'.
Bibliography : Cal. State Min. Bur. Register of Mines, Nevada County. U. S.
Geol. Survey Folio 29, Grass Valley Special.

The Grant patented fractional claim of 10 acres lies north of the Idaho-Maryland.

In 1898 an inclined shaft was sunk to a depth of 200' and 140' of drifting was done on the vein, which is said to be four feet in width. No ore was encountered and the mine has been idle since.

**Grass Valley Consolidated Gold Mines.** (See Allison Ranch Mine.)

**Great Britain.** (See North Star.)

**Great Eastern.** (See Syndicate Consolidated.)

**Green Mountain.** (See Sultana.)

**Grey Eagle Claim** (Sunny South). Owner, R. A. Campbell, 1014 Sixteenth street, Oakland, California.

> Location: Meadow Lake District, Sec. 27, T. 18 N., R. 13 E., 10 miles by trail from Cisco on S. P. R. R. Elevation 6900'.

**Grey Eagle.** (See Vulcan.)

**Grizzly Ridge Claim.** Owner, Grizzly Ridge Mining Company, North Columbia.

> Location: North Columbia Mining District.
> Bibliography: Cal. State Min. Bur. Bull. 50, page 204. U. S. Geol. Survey Folio 66, Colfax.

The owner reports an ore content as follows: gold $40, copper 1¼%, silver 17 ozs.

**Hartery Mine.** Owners, Ed. McLaughlin, San Jose; J. G. Loutzenheiser, Grass Valley.

> Location: Grass Valley Mining District, Sec. 2, T. 15 N., R. 8 E., 2 miles south of Grass Valley. Elevation 2300'.
> Bibliography: Cal. State Min. Bur. Rept. XIII, page 248. U. S. Geol. Survey Folio 29, Grass Valley Special.

This property, consisting of 12 acres, controls about 900' along two parallel veins which are the southern continuation of the Omaha-Lone Jack-Homeward Bound vein system. The veins average between 12" and 15" in width, strike N. 50° W. and dip 30° W. On the surface they are about 200' apart; the one to the westward is an extension of the Omaha vein which was so extensively worked in the adjoining Empire West group of mines. The veins occur in the granodiorite and were discovered and worked in the early 50's. From 1890–1893 the mine was operated by the Hartery Consolidated Mining Company, and $59,000 was produced during that period.

The west or Omaha vein has been developed to a depth of 600' by means of an inclined shaft. Levels were driven maximum distances of 300' north and 400' south from points 368', 508' and 600' below the collar of the shaft. A drain tunnel from Wolf Creek 1200' in length intersects the shaft at a depth of 250 feet. Some very rich specimens of free gold were taken from this vein and the ore is said to have averaged $30 per ton. During the operations of the Hartery Consolidated Company, an inclined shaft was sunk on the east vein to a depth of 500', but practically no exploratory work was done north or south of the shaft. The total production of the mine is given as about $350,000.

**Hermosa.** (See North Star.) Now part of North Star property.

**Heuston Hill.** (See Sultana.)

**Highgrade Claim.** Owner, John Clarke, Cisco, California. Bonded to Samuel Newhouse, and subbonded to third parties.

Location: Meadow Lake Mining District.
Bibliography: U. S. Geol. Survey Folio 39, Truckee, and Folio 66, Colfax.

A small force of men was employed in the spring of 1915, preparing for development work and hauling machinery and supplies. A compressor, small mill tramway, and gas engine were installed and a payment of $15,000 is said to have been made. The machinery was attached for debt in January, 1918, but when the property was visited by the sheriff most of the equipment was gone. Idle in 1918.

**Home.** (See Champion.)

**Homeward Bound.** (See Empire West Mines.)

**Idaho-Maryland Mine.** Owner, Gold Point Consolidated Mines, Inc.,* 1007 Crocker Building, San Francisco.

Location: Grass Valley, Secs. 25 and 26, T. 16 N., R. 8 E., 1 mile east of Grass Valley. Elevation 2540'.
Bibliography: Cal. State Min. Bur. Repts. VI, pt. II, page 45; VIII, page 425; X, page 372; XI, page 271; XII, page 192; XIII, page 251. U. S. Geol. Survey 17th Ann. Rept., pt. II, Gold Quartz Veins of Nevada City and Grass Valley Districts (Lindgren). U. S. Geol. Survey Folio 29, Grass Valley Special; U. S. Geol. Survey. Ross Browne and Raymond Reports on Mineral Resources West of the Rocky Mts., 1867–1877.

The Eureka-Idaho-Maryland vein was marked by a conspicuous outcrop of white quartz on the Eureka claim, which was located in 1851. Owing to the fact that this vein was very low grade near the surface, it was worked only in a desultory manner, at a loss, until rich ore was encountered in 1863 at a depth of 100 feet. Eastward of the Eureka claim in the western portion of the Idaho ground, the vein is less marked and it can not be traced on the surface east of the Idaho shaft, although the great Idaho-Maryland ore shoot has been worked for a distance of 3500' east of this shaft.

The Idaho mine was consequently not located until 1863 when the rich strike was made in the adjoining Eureka ground. Little work was done until 1865 when a vertical shaft was sunk to a depth of 120' without finding ore. In 1867 the Idaho Quartz Mining Company was formed with Edward Coleman as President, and the great pay shoot was encountered at a depth of 300 feet. The mine was thereafter worked continuously until 1893, and during that period this wonderful pay shoot was stoped for a distance of 3500' along the strike of the vein to the eastern boundary of the Idaho ground. Litigation then ensued and in February, 1893, the Maryland Company, which owned the adjoining land to the east, acquired the Idaho property by the payment of $85,000. The Idaho hoisting plant was destroyed by fire

*Since the above was written this property has passed into the possession of the Idaho-Maryland Mines Company, Hobart Bldg., San Francisco.

Photo No. 27.   The Idaho-Maryland Mine, from the Eureka Shaft.   Looking east along the strike of the vein.

in January, 1894, resulting in the filling of the lower levels which in the Canyon shaft territory have never since been reopened below the 1600' level. The mine was operated under the management of S. P. Dorsey until 1901, when it was closed on account of the bad condition of the workings, the distance the ore had to be transported, and the number of times it had to be transferred before reaching the surface. In 1903 the mine was bonded to the Idaho-Maryland Development Company and was worked by them until October, 1914. During this period of eleven years, the Company only succeeded in reopening the mine to the 1000' level and the ore milled came, for the most part, from the old stopes and pillars which were left behind during previous operations. The mine is at present (1918) idle and inaccessible on account of water.

The records of the production of this property under the different regimes are complete and authentic.

The gross production and dividends paid are as follows:

**Copy of Annual Report Made by Edward Coleman, President of the Idaho Quartz Mining Company.**

| Year | Tonnage | Production | Dividends |
|------|---------|-----------|-----------|
| 1868 | 763 | $45,534 80 | |
| 1869 | 9,489 | 308,208 65 | $170,500 00 |
| 1870 | 9,782 | 189,963 57 | 37,200 00 |
| 1871 | 11,133 | 395,355 63 | 232,500 00 |
| 1872 | | 400,465 42 | 162,750 00 |
| 1873 | 28,825 | 1,024,591 89 | 682,000 00 |
| 1874 | 28,401 | 664,811 20 | 317,750 00 |
| 1875 | 28,103 | 495,569 50 | 172,050 00 |
| 1876 | 29,720 | 562,274 06 | 255,750 00 |
| 1877 | 29,250 | 630,143 69 | 240,250 00 |
| 1878 | 33,883 | 596,850 33 | 263,500 00 |
| 1879 | 32,370 | 499,379 61 | 168,950 00 |
| 1880 | 11,611 | 226,078 95 | 127,100 0 |
| 1881 | 27,540 | 612,588 84 | 271,250 00 |
| 1882 | 27,540 | 568,572 45 | 263,500 00 |
| 1883 | 28,572 | 364,599 85 | [1]34,500 00 |
| 1884 | 31,143 | 561,895 49 | 271,250 00 |
| 1885 | 30,518 | 370,197 71 | 99,200 00 |
| 1886 | 29,244 | 547,569 82 | 263,500 00 |
| 1887 | 26,686 | 492,638 07 | 233,600 00 |
| 1888 | 26,664 | 603,634 92 | 525,500 00 |
| 1889 | 21,448 | 407,385 25 | 178,250 00 |
| 1890 | 20,321 | 268,904 35 | 52,700 00 |
| 1891 | 16,759 | 314,087 06 | 10,250 00 |
| 1892 | 16,500 | 248,270 57 | 57,450 00 |
| 1892 | January | 40,904 83 | [2]23,783 58 |
| Cash received from the Maryland Gold Quartz Mining Co. | | | 85,000 00 |
| Totals | 567,029 | $11,470,573 90 | $5,093,433 18 |

Average per ton, $20.23.
[1]New equipment. [2]Balance.

The secretary's final report to the company gave a gross production of $11,873,553.81 and dividends aggregating $5,017,063.58.

**Dorsey Production.**

| | |
|---|---:|
| 1893—11 months | $258,220 34 |
| 1894 | 193,182 52 |
| 1895 | 247,600 86 |
| 1896 | 197,239 78 |
| 1897 | 147,646 16 |
| 1898 | 93,242 20 |
| 1899 | 68,344 91 |
| 1900 | 31,503 96 |
| 1901 | 9,040 40 |
| Total | $1,246,020 13 |

From 1904 to October, 1914, the Idaho-Maryland records show a production of $236,613 for the company account, to which should be added $75,000 for tribute account.

The Idaho-Maryland mine has therefore produced, from 1867 to 1915, $13,431,187 and has paid from this amount $5,017,083.58 in dividends. The average value of the ore milled was about $20 per ton and the total costs of mining and milling have ranged from $10 to $15 per ton, as compared with the present day working costs of $5 per ton at the North Star and Empire mines.

The adjoining Eureka mine produced from 1865 to 1877 approximately $5,700,000 and paid $2,054,000 in dividends. The wonderful Eureka-Idaho-Maryland pay-shoot has therefore produced $19,131,187 and has paid $7,071,084 in dividends.

As has previously been stated, the only prominent outcrop of the Eureka-Idaho vein occurred on the Eureka claim, where a vein of massive white quartz is found in the serpentine. In the vicinity of the old Eureka shaft east of the Idaho shaft there are indications of decomposed and altered diabase and schists. It is said that there were two parallel veins in the Eureka ground separated by a dike or horse of diabase from 2' to 30' in thickness, with the main vein lying between the serpentine and the diabase. In the Idaho-Maryland mine, however, the vein occurs at or near the contact of the serpentine and diabase; the latter forms the hanging wall of the vein and is, in many cases, so altered and decomposed that it is difficult to determine the original rock. One of the few crosscuts driven in the Idaho-Maryland ground was started from the 700' level in the Idaho shaft and was extended in a southerly direction a distance of 800' to develop the South Idaho vein, which lies 1200' south and parallel to the Idaho vein. For the first 25' the crosscut was in diabase, but at that distance gabbro was encountered and traversed for a distance of 200', where diabase again appeared and continued to the face of the tunnel. The diabase in the face of the crosscut was cut by numerous small seams, but failed to show any distinct veins of mineralization; it is reported, however, that a 'soft formation' was encountered and that the crosscut was discontinued for fear of striking a flow of water.

The serpentine of the footwall is altered near the vein to a dolomitic rock sometimes colored green by mariposite. This occurrence is characteristic of most of the veins which are associated with the serpentine areas traversing Nevada, Sierra and Plumas counties. The foot-wall serpentine of the Idaho-Maryland vein when exposed by the development had the property of increasing in volume due to the absorption of moisture, and is locally known as 'swelling ground.' During the operation of the mine considerable difficulty was experienced in keeping the workings timbered and open and this was one of the main reasons given for closing the mine.

The Eureka-Idaho-Maryland vein has a general strike of N. 77° W. and a dip varying from 50° to 80° E., averaging about 70°. Underground within the ore-shoot the vein averaged from $2\frac{1}{2}'$ to 3', but beyond the confines of the pay shoot it frequently narrows down to a mere veinlet or seam, which is characteristic of other veins in the Grass Valley district. In some portions of the ledge in the Eureka and Idaho ground, the veins are of average width, but the ore is low grade. Within the pay shoot the vein sometimes attained a maximum width of 8 feet. The form of the pay shoot is shown in the accompanying figure (No. —). The ore consists of quartz and the country rock contains small quartz stringers which in many cases carry coarse free gold. The percentage of sulphides in the ore is comparatively small, averaging not more than 2% and consisting for the most part of pyrite, together with chalcopyrite, galena and a very small percentage of zinc-blende. The principal value lies in the free gold content, which as a rule, is finely disseminated throughout the quartz, but as is the case in the Empire and North Star mines, rich 'bunches' of 'specimen ore' are often found, many thousands of dollars being taken out from a comparatively small area.

The ore milled from the Eureka-Idaho pay shoot has varied from $10 to $60 per ton, the latter value having been taken from the Eureka ground; it is estimated that the ore from the mile of workings on this pay shoot has averaged over $20 per ton. The following is quoted from the sworn statement of a number of the most prominent mining men of Grass Valley who inspected the Idaho mine a few days prior to its closure:

"The ledge we saw in the 1600' east incline or Dorsey shaft, and the levels from this shaft, is a continuation of the Eureka-Idaho-Maryland pay shoot of quartz on its eastward and downward strike, in the regular fissure and is of good size, viz, two to four feet, and the ore appears to be of good quality."

It is further reported that some of the quartz in the faces mentioned showed free gold.

According to Lindgren, this pay shoot is "one of the most remarkable known to vein geology" and "there is no reason why it should not continue for a long distance eastward, provided the vein does not enter the serpentine; if it does, the probability is that the vein will be found to split up in stringers."

It should be remarked, however, that from a study of the maps of the workings, it appears that the easterly end of the shoot is in a more disturbed area than had previously been met with in the 5000' of development and stoping to the westward. It is reported on good authority that a small ore shoot was found on the 1700' and 1800' levels from the Cañon shaft, and considerable ore was stoped; the vein, however, was considered too small to warrant further attention. The workings on this shoot have never been accessible since the flooding of the lower levels at the end of the Coleman regime in 1893. In view of the development, in the other large mines of the Grass Valley district, of numerous disconnected ore-bodies often hundreds of feet apart, it appears that this Cañon shaft ore shoot offers an excellent opportunity for the development of another ore body entirely independent of the Eureka-Idaho-Maryland pay shoot; nor can it be said that these two orebodies exhaust the possibilities of the property, for there is every reason and geological precedent to expect that aggressive and systematic development work will open up other pay shoots on the known lode and there is also the possibility that other parallel or diverging veins will be discovered.

The number and extent of the various shafts, drifts, crosscuts and raises and the stoped area are clearly shown in the accompanying maps and sections and for that reason no verbal description will be attempted.

In regard to the best methods for reopening and developing the extensions of the known ore shoots a number of suggestions have been advanced. At the time the property was bonded by the Idaho-Maryland Development Company, the plan was to reopen the old Idaho shaft, and continue the sinking of this shaft 1000' or to a total depth of 2000' and then to drift eastward, encountering the Cañon shaft shoot at a distance of 1500'. This drift was then to be continued a distance of 2000' east, to the bottom of the Dorsey shaft. So many unexpected difficulties and delays were encountered in reopening the Idaho shaft and the excessive cost of opening and retimbering were so great that the company made no attempt to continue the work of sinking the Idaho shaft or to reopen the old Cañon incline, practically confining their attention to the vain endeavor of discovering ore in the worked-out ground above the 1000' level. In the opinion of those familiar with the underground conditions in the past, all the workings below the 1000' level on the contact of the swelling serpentine and altered diabase and schists are closed, and it is further contended with apparent correctness that it would cost more to reopen the old shafts and drifts than to open up the known ore-bodies by entirely new work, which would also explore undeveloped portions of the lode. Former superintendent Nye has suggested that the best plan to reopen the mine would be to continue the driving of the 1000' level eastward a distance of 2000' and then to sink a winze in the hanging-wall of the vein. This winze would have to be sunk a distance of approximately 1000' before reaching the lower levels of the Dorsey shaft. The plan further contemplated the connection of the winze with the surface, thus making a main working shaft 2000' in depth. Another plan was to sink a vertical shaft in the diabase hanging wall, which would cut the Idaho-Maryland vein at a vertical depth of from 3500 to 4000' or from 4000 to 4500' on the dip of the vein. This shaft would be about 800' in the hanging wall of the Idaho-Maryland vein at 2000'; the depth of the lowest working of the Cañon and Dorsey inclines; a crosscut would therefore have to be driven north a distance of 800' to intersect the old workings. This plan has many advantages to recommend its adoption; it would give a large vertical shaft in diabase needing little expenditure for up-keep as compared with a shaft at or near the serpentine contact; the shaft could be centrally located to open up not only the Cañon shaft and Dorsey shoots, but also to prospect the numerous veins in the hanging wall of the Idaho lode. There is no doubt but that the Idaho-Maryland territory together with the South Idaho and Gold Point properties offer one of the few opportunities yet remaining in the Grass Valley district which has not been acquired by the big operating companies.

If the combined properties are efficiently operated by a company with sufficient capital to carry forward an aggressive development campaign there is every reason to expect that these mines would again become, as in the past, the premier gold-producing property of the Grass Valley district, outrivaling the Empire and North Star mines, which are each producing gold at the rate of over $1,000,000 per annum.

The Idaho-Maryland is equipped with a 20-stamp electrically-driven mill; a flat rope water-power hoist; a 14" Cornish pump operated by electricity, handling from 250 gallons a minute during ten months of the year, to 500 gallons a minute, during a short period in the rainy season. The buildings housing this equipment and the equipment itself are not in good condition and while they may suffice for the unwatering and preliminary reopening of the mine, an entirely new plant and equipment will be required before mining can be conducted upon a scale necessary for successful operation.

**Illinois.** (See Empire West Mines.)

**Illinois.** (See Rocky Glen.)

**Independence.** (See Canada Hill.)

**Independence Claim.** Owner, D. Fricot, Angels Camp, Calaveras County, California.

> Location: Grass Valley Mining District, Sec. 25, T. 16 N., R. 8 E., 2 miles east of Grass Valley. Elevation 2700'.
> Bibliography: Cal. State Min. Bur. Rept. XIII, page 248. U. S. Geol. Survey Folio 29, Grass Valley Special.

The Independence claim lies southeast of the Idaho-Maryland and north of the Brunswick property. In 1896 a small inclined shaft was sunk in the serpentine to a depth of 800 feet. At this depth a crosscut was driven south toward the contact of the serpentine and the Brunswick amphibolite schist. A vein is said to have been encountered and a drift was driven thereon. No ore, however, was developed. The property has remained idle for the past 15 years.

**Indiana.** (See Ben Franklin.)

**Inkmarque Group** (Inkerman & Lemarque). Owners, North Star Mines Company.

> Location: Grass Valley Mining District, Sec. 3, T. 15 N., R. 8 E., 2 miles south of Grass Valley. Elevation 2400'.
> Bibliography: Cal. State Min. Bur. Rept. XIII, page 249. U. S. Geol. Survey Folio 29, Grass Valley Special Map.

The claims comprising this property are the Inkerman, Lamarque, Kemp and White Oak, about fifty acres of patented mineral land. The Inkerman-Lamarque vein belongs to the North Star system of east-striking, north-dipping fissures. The vein was worked to a shallow depth in early days and was reopened in 1904 by the present owners. A shaft was sunk on the vein, dipping 20° N. to a depth of 500 feet, at which point the incline broke into the old workings of the North Star mine, proving that the two veins united at this depth. This is but one of the many examples of the complexity of the vein structure in the Grass Valley district. After it had been proven that the Inkmarque was not an independent footwall vein, but united with the North Star fissure, work was discontinued and the equipment was removed from the property. On the White Oak claim there is a vein which strikes N. 15° E. and dips to the west, but very little work has been done on this vein.

(The property has since been acquired by the North Star Mines Co.)

**I. X. L.** (See North Star.)

**I. X. L. Claim.** Owner, A. H. Woodworth, 319 Van Ness avenue, San Francisco.

> Location: Washington Mining District, Sec. 28, T. 18 N., R. 11 E., 4 miles northwest of Washington.
> Bibliography: U. S. Geol. Survey Folio 66, Colfax.

The I. X. L. vein occurs in slate and strikes north. It dips 70°–80° E. The Gaston vein is parallel to it and lies a mile east at the contact of the slate and granite. No work has been done on the claim for many years. The Bonanza and Lucky Boy patented claims situated about one mile southeast of the I. X. L. and adjoining the Gaston property on the south, are controlled by the same owner.

**Keller Mine.** Owner, F. A. Garbutt, Olive street, Los Angeles.

Location: Graniteville Mining District, Sec. 8, T. 18 N., R. 11 E., 2 miles west of Graniteville. Elevation 3900'.
Bibliography: U. S. Geol. Survey Folio 66, Colfax.

Property consists of the Montana placer claim of 140 acres and three patented claims of twenty acres each, known as the Mayflower, Baltimore and Dewey, located in 1890. These claims cover a length of 4500' along the northern extension of the Republic-National-Culberson lode. The vein can be traced for a number of miles and has been worked at different points.

The Keller vein occurs in slate, strikes N. 18° E., dips 60° E., and varies in width from 1' to 18' with an average of 8 feet. It has been developed by means of a 1400' crosscut adit which encountered the vein at a depth of 600 feet below the surface. From the end of this crosscut adit, drifts were driven on the vein 300' north and 300' south, and an ore shoot 275' in length was opened. This pay-shoot averaged 8' in width and the gold recovered from the ore averaged $4.50 per ton. The tailings were not sampled and the loss therein is unknown. A raise was started from the lower level in the shoot to connect with a winze which had been sunk 120' below the upper tunnel, but the work was discontinued before the connection was made. About 1600 tons of ore was stoped and the yield is said to have been $7500.

Equipment consists of boarding house, bunk houses, and blacksmith shop. The mine was last worked in 1913.

**Kenosha Mine** (Seven Thirty). (See Alcalde Gold Mines Company.)

**Kentucky.** (See Golden Gate.)

**Lemarque.** (See Inkmarque.)

**Larimer.** (See North Star.)

**Last Chance Mine.** Owner, S. L. Parsons Estate, Washington, Nevada County, California.

Location: Washington Mining District, Sec. 14, T. 17 N., R. 11 E., 6 miles southeast of Washington. Elevation 5300'.
Bibliography: Cal. State Min. Bur. Rept. XIII, page 250. U. S. Geol. Survey Folio 66, Colfax.

The vein which has been explored on the Last Chance claim is a southern continuation of the Eagle Bird fissure system. It is not as wide as that worked on the Eagle Bird property, but some high grade rock is said to have been produced in the past. The vein strikes north and dips 60° E. The mine was not being worked in 1914.

**Last Chance.** (See under Copper.)

**Le Compton Mine.** Owner, James J. Ott.

Location: Nevada City Mining District, Sec. 9, T. 16 N., R. 9 E., 3 miles east of Nevada City. Elevation 3000'.
Bibliography: U. S. Geol. Survey Folio 29, Nevada City Special.

The Le Compton vein was first worked from 1857 to 1867 and produced during that time, supposedly $250,000. Closed in 1867 on account of the flooding of the incline, no work was done until the property was reopened in recent years and worked at intervals. The vein, which outcrops on the south side of Deer Creek, occurs in the granodiorite, striking east and dipping 40° N. It has been developed by an inclined shaft. Idle in 1918.

**Le Duc Mine.** Owner, A. Le Duc, Grass Valley. Property under bond to Le Duc Gold Mining Company; C. F. White, president, San Francisco; W. G. Thomas, secretary, Grass Valley.

Location: Grass Valley Mining District, Sec. 6, T. 15 N., R. 9 E., 2 miles southeast of Grass Valley. Elevation 2200'.
Bibliography: U. S. Geol. Survey Folios 18 and 29.

The Le Duc Gold Mining Company has been operating this property, consisting of 122 acres of agricultural land, for six years previous to 1914 under a ten-year bond. The northern half of the ground is capped with from 50' to 100' of andesite, tuff and breccia. Under this capping there is a tertiary channel which has been worked at a number of points on the western edge of the lava cap. The Alta Channel lying northwest of Grass Valley is probably the western continuation of this same stream. Gravel has been encountered in a number of wells which have been sunk through the lava and also by a raise from the end of the Le Duc tunnel.

On the southern portion of the property there are five quartz veins, but these are covered toward the north by the andesite. Two of the veins have been explored to a limited extent by the Le Duc adit, which was driven in a northwesterly direction from a point 600' south of the Grass Valley-Colfax road. The tunnel was turned 500 feet from the portal and continued in a more northerly direction for 400' where a vein, striking N. 40° W. was encountered. It followed this vein, which varied in width from 2" to 4' for a distance of 300 feet. Samples taken therefrom are said to have ranged in value from $10 to $58 per ton. This vein occurs at the contact of the slate and diabase, and after following the ledge for 300', the main adit was driven N. 75° W. for a distance of 600' and then S. 60° W. for 400 feet. A short drift was extended northward from the bend in the adit and a 290' raise was put through to the surface. This raise encountered gravel capped with about 60 feet of lava. From the point where the main adit left the vein, a drift has been driven 350' in a northerly direction. This drift will be driven 200' farther, and a raise will be made to find the channel.

Lying about 100' east of the vein which was developed by the main adit is another vein which is said to occur at the contact of the diabase and serpentine. A short drift has been driven on this vein from the adit level, but no ore-body has been found. The veins on this property warrant further development. The only equipment on the property consists of a blacksmith shop.

At an annual meeting of the directors and stockholders, held in January, 1916, it was decided to continue development and operation. More than 3400' of tunnels have been driven, which have cut several ledges, but the pay shoot or ore-body has not been disclosed. The company has spent about $40,000 for development and expects to continue until successful.

(No activity was reported for this property in the fall of 1918.)

**Liberty Hill.** (See Empire and Pennsylvania.)

**Little Done.** (See Alpine.)

**Lone Jack.** (See Empire West Mines.)

**Lone Star.** (See Murchie.)

**Lotzen Ranch.** (See under Copper.)

**Lucky.** (See Union Hill.)

**Madison Hill.** (See Sultana.)

**Mammoth.** (See National.) Owner, P. Thautphaus.

**Mammoth Gold Copper Mine.** (See under Copper.)

**Mammoth Group.** Owner, _____

Location: Meadow Lake Mining District, Sec. 4, T. 17 N., R. 13 E., 9 miles southeast to Cisco (S. P. R. R.), good mountain wagon road.
Bibliography: Cal. State Min. Bur. Bull. 50, page 204. U. S. Geol. Survey Folio 39, Truckee.

The locations which make up the Mammoth Group of claims cover a total area of 80 acres.

Development work consists of a 285' vertical shaft and 75' of drifts in ore. There are also two adits, No. 8 adit being 75' long, and No. 2 adit being 200 feet. The vein walls are porphyry, and they strike northwest and dip to the south. The gossan is iron-stained porphyry. The reported value of the ore is gold $15, copper 4% and a trace of silver.

Equipment consists of machinery for a ten-stamp mill, but this is not in place.

**Manhattan.** (See Goodall.)

**Mary Ann Claim.**  Owner, H. M. Foster, Grass Valley.

Location: Grass Valley Mining District, Sec. 3, T. 15 N., R. 8 E., 2½ miles south of Grass Valley. Elevation 2400'.

The Mary Ann claim of 20 acres covers rights to the Mary Ann-Phoenix vein, a part of the Allison Ranch-Omaha fissure system. This vein strikes north and dips 40° W., paralleling the Allison Ranch veins, which lie 2500' to the east. Practically no development work has been done on the vein since the Phoenix shaft was sunk to a depth of 100 feet in the early 70's. At that time it was reported that ore averaging $20 per ton was produced.

**Massachusetts Claim.**  Bonded to H. O. Secrist, Nevada City.

Location: Nevada City Mining District, Sec. 18, T. 16 N., R. 9 E., 1½ miles south of Nevada City.
Bibliography: U. S. Geol. Survey Folio 29, Nevada City, and Folio 18, Smartsville.

The Massachusetts quartz claim at Gold Flat, between the Phoenix and Pittsburg mines, was taken under bond by H. O. Secrist of Nevada City, in March, 1916. Good ore has been taken from the upper workings, and the ledge is stated to average around $50 per ton. The oreshoot contains free-milling quartz assaying $40 to $150 per ton.

(No record of recent work was to be had in September, 1918, and the mine was idle.)

**Massachusetts Hill.**  (See North Star.)

**Mayflower Mine.**  Owner, Mayflower Quartz and Channel Mining Company, Nevada City, California.

Location: Nevada City Mining District, Secs. 17 and 20, T. 16 N., R. 9 E.
Bibliography: Cal. State Min. Bur. Repts. VIII, page 453; XI, page 295; XII, page 192; XIII, page 252. U. S. Geol. Survey Folio 29, Banner Hill District.

The Mayflower placer mine of 149 acres was located in the early days and the channel of that name was worked by hydraulicking. Subsequently, auriferous quartz veins were discovered and in 1879 the property came into the possession of W. H. Martin. The different veins were developed and worked until 1901, but no exploration of any magnitude has been done from 1901 to 1915. The total production of the mine is said to be "several hundred thousands of dollars."

There are eight veins on the property, known as the Grant, which strikes N. 15° W. and dips 45° E.; the Beckman, which strikes E. and dips from 0° to 30° S.; the Mayflower, which strikes N. 60° W. and dips 20° S.; the Floyd, which strikes N. 10° W. and dips 45° E., and the Butterfly, North Star and Big Blue, which strike from S. 70° E. to E. and have a vertical dip. Both the Floyd and Beckman veins are faulted by the Butterfly, North Star and Big Blue veins. The movement, however, in all cases has not been more than 20 feet.

The greater part of the development has been confined to the Beckman and Grant veins and the other veins have only been worked near their intersection with these.  A 300' inclined shaft was sunk on the Grant near the point where it crosses the contact of the slate and granodiorite.  The vein only averaged about 15" in width, but rich ore is reported to have been stoped; the quartz contains, besides free gold, 3% of sulphides, pyrite, arsenopyrite, galena and zinc blende.

The Beckman vein was worked to a depth of 215' and for a length of 1500' by means of an 800' adit.  A shaft was also sunk, developing the ledge to a depth of 700 feet.  The 600' or lowest level was driven 1000' east of the shaft.  No information could be obtained regarding the extent or value of ore at this depth.  The showing so far made warrants further development, but a consolidation of this property with adjoining properties would be advisable before exploiting this part of the district on a large scale.

The 20-stamp mill was placed in commission in the latter part of November, 1915, and was said to be running on quartz from old levels. A vertical shaft has been sunk 180' and drifting has started.  C. E. Brackett is superintendent.

(The property was idle in September, 1918, and no record of results obtained in 1915 was obtainable.)

**Menlo Gold Mining Company.**  (See Empire West Mines.)

**Merrifield.**  (See Champion.)

**Merrimac Mine** (Defiance).  Owner, Merrimac Mining and Milling Company; Col. Geo. Stone, agent, 200 Cherry street, San Francisco.

Location: Nevada City Mining District, Sec. 24, T. 16 N., R. 8 E., 2 miles south of Nevada City.
Bibliography: Cal. State Min. Bur. Repts. XI, page 279; XIII, page 253.  U. S. Geol. Survey Folio 29, Nevada City Special.

In the early '90's, a 385' inclined shaft was sunk on the vein which strikes east and dips 42° N.  The vein occurs in the black Mariposa slates, and is said to be from 18" to 3' in width.  Drifts have been driven a maximum distance of 350' east of the shaft and the ore shoot developed is reported to have been 300 feet long.  The mine has been idle for many years.  The nearest operating mine is the Pittsburg, which lies 3000' northeast.

**Metropolitan Group.**  Owner, Jennie Reynolds, Grass Valley, Nevada County, California.

Location: North Bloomfield Mining District, Sec. 15, T. 18 N., R. 10 E., 5 miles northeast of North Bloomfield.  Elevation 3800'.
Bibliography: Cal. State Min. Bur. Rept. XIII, page 253.  U. S. Geol. Survey Folio 66, Colfax.

There are three patented claims and a millsite in this group, covering 4500' along the lode.  The vein, which has been developed by means of an adit over 1000' in length, occurs in the same area of amphibolite as the famous mines of the Alleghany district, lying about three miles to the north. The mineralogical characteristics of the ore are also similar to the Alleghany mines as the rich 'bunches' of ore are associated with arsenopyrite.

The vein varies in width from 2' to 4'. Its strike is N. 50° W. and the dip 68° SW. The mine was not being operated in 1914.

**Mineral Hill.** (See under Copper.)

**Mistletoe Claim.** Owner, C. J. Graham, Nevada City.

> Location: Rough and Ready Mining District, Sec. 23, T. 16 N., R. 7 E., 1 mile west of Rough and Ready. Elevation 1800'.
> Bibliography: Cal. State Min. Bur. Repts. XII, page 193; XIII, page 253. U. S. Geol. Survey Folio 18, Smartsville.

The vein, which strikes N. 70° W. and dips 60° NE., occurs in granodiorite. It has been only superficially developed, but is said to average 2' in width. One patented claim known as the Mistletoe comprises the property.

**Mohigan Group** (Gold Flat). Geo. B. Finnegan, agent, Nevada City.

> Location: Nevada City Mining District, Sec. 13, T. 16 N., R. 8 E., 1½ miles south of Nevada City. Elevation 2740'.
> Bibliography: Cal. State Min. Bur. Rept. XII, page 189.

This prospect includes the Gold Flat and Mohigan claims, covering 2500' along the Mohigan vein which strikes E. and dips 38° S. The vein averages about one foot in width and has been worked for a length of 800' by an inclined shaft, 150' deep. The ore stoped is said to have averaged $30 per ton. The mine has not been worked for a number of years.

**Montana-Willow Valley Group** (Willow Valley Mining Company). Owner, A. W. Hogue, Nevada City, California.

> Location: Nevada City Mining District, Sec. 4, T. 16 N., R. 9 E., 2½ miles east of Nevada City. Elevation 3000'.
> Bibliography: Cal. State Min. Bur. Rept. XIII, pages 254, 267. U. S. Geol. Survey Folio 29, Banner District.

The Willow Valley claim (1500' x 600'), the Montana (800' x 200') and the North Montana (945' x 475') comprise this property. The total area is about 32 acres, and the claims control 2450' along the Montana-Willow Valley vein system. The Montana vein which was first worked in the early '50's, strikes N. 50° E. and dips 22° NW. In 1870 an inclined shaft was sunk on the vein to a depth of 350' and ore was stoped from the 250' level to the surface. The length of the shoot is said to have been 400' and the average width one foot. It is reported that the 6000 tons of ore yielded an average of $30 per ton. The New York vein, lying 250' in the hanging wall and paralleling the Montana vein, was also worked in early days; it is said to have produced $80,000 from 4000 tons of ore in a depth of 200 feet. In 1907 and 1908 a drain tunnel, 527' in length, was driven from a branch of Mosquito creek, south to the Montana shaft, which it intersected at the 200' level. This tunnel cut the New York vein at a distance of 67' from the portal. A drift was driven on the vein 100' in a northeasterly direction and some ore was stoped. Another parallel vein, known as the Middle vein, lying midway between the New York and Montana, was prospected by means of a 100' drift. This vein was 18" in width and 100 tons of ore was stoped which is said to have averaged $9 per ton.

In 1909 a hoist was installed and the Montana shaft was sunk to a depth of 830' on the vein. On the 400' level considerable exploratory work by means of drifts, crosscuts and raises was done on the different veins. For the first few hundred feet, the 400' northeast drift was driven on a foot-wall vein, then it was driven in a northerly direction for 110' at which point the Montana vein was intersected. This drift was continued 270' northeast into the Willow Valley claim. A crosscut driven northeast into the hanging wall, encountered two veins, one at a distance of 40' and the other at 150 feet. From raises, stopes and drifts on the 400' level, 750 tons of ore were milled which averaged $9.18 per ton.

A crosscut was driven in a northerly direction, until it encountered the New York vein with which the Middle vein is supposed to have formed a junction and a drift was driven on this vein 320' northeast. The vein averaged 2' in width, and is said to have assayed from $4 to $10 per ton, but no ore was stoped. The sinking of the shaft is the only exploratory work done below the 600' level. The New York and Montana veins are thought to unite at a depth of 800', forming a vein whose average width is 4'.

The Willow Valley vein was located in 1865 and worked at intervals, the last work being done in 1904. The vein has been developed to a depth of about 150 feet. The total production is estimated to have been $50,000 from 3000 tons of ore.

Equipment on the property consists of an electric hoist capable of working to a depth of 2000'; 10" x 12" Sullivan air compressor operated by a 50-horsepower G. E. motor, complete blacksmith shop, transformer house with 3-35-horsepower transformers; a rock-breaker, 5-stamp mill, and Johnson concentrator, housed in a corrugated iron building. Electric power is now used throughout at a cost of 0.72 cent per K. W. H. The property was idle in 1914.

The Willow Valley Mines Company installed an electric pump with a capacity of 500 gals. per min. and in 1915 were reported to be preparing to sink from the 830' point to a depth of 1200'. Arthur Hogue, Superintendent. (Idle in September, 1918.)

**Mountaineer Mine.** Owner, Mountaineer Mines Company; P. and J. Bender, and F. J. J. Sloat, all of Morgan Hill, Santa Clara County, California.

Location : Nevada City Mining District, Secs. 12 and 13, T. 16 N., R. 8 E., 1 mile west of Nevada City. Elevation 2400'.
Bibliography : Cal. State Min. Bur. Repts. X, page 384 ; XI, page 287 ; XII, page 193 ; XIII, page 254. U. S. Geol. Survey Folio 29, Nevada City Special.

The Mountaineer property consists of 300 acres of lode, placer and agricultural patented land lying west and south of the Nevada City townsite. The Champion mines group owned by the North Star Mines Company, adjoins the Mountaineer holdings on the west. The Mountaineer and Black Prince veins belong to the Champion-Merrifield system of north and south-striking, east-dipping veins, but there has been less movement along the Mountaineer fissure and the veins are therefore smaller than those of the famous Champion group. The Mountaineer vein, near the southern end lies about 500' east of the Merrifield vein, but toward the north the veins gradually diverge until a mile north of the Providence-Merrifield shaft the veins are 1500' apart. The strike of the Mountaineer vein is N. 18° E. and the dip averages 37° E., although many local irregularities occur. The vein averages about one foot in width, but at intervals it swells to a maximum width of ten feet; in the enlarged portions, however, the ore is usually of low grade. The walls of the vein are composed of unaltered granodiorite. The ore is ribbon quartz, carrying free gold, and 3% to 4% of sulphides, consisting chiefly of pyrite with smaller amounts of chalcopyrite, sphalerite and galena. The sulphides carry from $100 to $200 per ton in gold and silver. The percentage of silver in the ore is larger than in most of the mines in the Nevada City and Grass Valley mining districts.

The Black Prince vein which lies 150' east and parallel to the Mountaineer lode is a large quartz-filled fissure interlinked with the Mountaineer by cross fissures and seams. The Black Prince is said to average from 2' to 3' in width, with a maximum of 12 feet. This vein also lies between granodiorite walls. No ore-shoots have been encountered and the massive white quartz is practically barren as is also the quartz between pay shoots in the Mountaineer vein.

The Mountaineer vein has been developed by means of a 2000' adit, driven northward on the vein from the north bank of Deer Creek. At a distance of 1000' from the portal, a shaft was sunk on the vein to a depth of 1200 feet.

Stations were cut 100', 200', 300', 400', 500', 600', 700', 900' and 1200' below the tunnel level and drifts were run a maximum distance of 1500' north and 2200' south of the shaft. The twelfth level was driven 1500' north on the vein, but no exploratory work was done south of the shaft on this level. The lowest working south of the shaft is the 900' level which is 1500' in length.

The development work opened five pay shoots in all. The main ore-shoot occurred 100' north of the shaft at the tunnel level and has a rake or pitch in the plane of the vein of 40° N. At the 400' level this vein pinched down to a small seam, but opened out again on the 500' level where the vein had an average width of 20" and the shoot was 160' in length. Below the 600' level the vein increased in size until a maximum width of 10 feet of practically barren quartz was reached. Sixty feet above the 1200' level this pay-shoot had again narrowed to a width of 30", and the ore is said to have averaged $11 per ton. Another ore-shoot occurred about 200' north of the main shoot, but this was only 15' in length and the ore averaged only 5" to 6" in width. Three pay shoots were opened at different points south of the shaft, but the ore for the most part was low grade.

A new shaft located about 300' north of the old Mountaineer shaft has been sunk on the Black Prince vein. This shaft has reached a depth of 400' on the vein, but little lateral development work has been done. It is equipped with a water-power hoist capable of sinking to a depth of 2000 feet. The Live Yankee, Summit, Dodo, Orleans and Orleans Extension, which also comprise part of the Mountaineer holdings, cover 5000' along the Fortuna-Orleans-Glencoe vein. This vein is one of the longest in the Nevada City district, being traceable from the Glencoe to the southern end of the Champion vein system, a distance of 2 miles. Only superficial exploratory work has been done on the vein which occurs in the narrow belt of siliceous clay-slates bordering the Nevada City granodiorite area on the south. The southern end of the Champion vein system also enters and apparently terminates in this same belt of Calaveras slates. The strike of the Orleans lode is N. 70° W. and the dip is 70° to 80° S. Two shafts have been sunk on this vein, the Fortuna, 50' west of the upper Grass Valley road, and the Orleans, 900' east of the lower Grass Valley road. Both of the shafts are only a few hundred feet in depth. An adit known as the Orleans tunnel located about 500' east of the railroad, has been driven for a short distance on the vein. This superficial development has failed to open up any orebodies, but in several places low-grade quartz with a considerable amount of sulphides has been encountered. The showing so far made, however, warrants further exploratory work.

The equipment on the Mountaineer property consists of a 10-stamp mill, a 200-horsepower Laidlaw-Dunn-Gordon compound compressor, blacksmith shop, tools, etc., and a water-power hoist, capacity 2000 feet. Water supplied by the Pacific Gas and Electric Company at 16¢ per inch, is used under an 800' head for power.

The total production of the Mountaineer mine is reported to have been from $2,000,000 to $3,000,000. It was last worked in 1913 in a small way by lessees.

F. J. J. Sloat of Morgan Hill is the principal owner (April, 1916). Idle in September, 1918.

**Mount Auburn Mine.** Owners, James H. White et al., Grass Valley, California.

> Location: Nevada City Mining District, Sec. 2, T. 16 N., R. 8 E., 2½ miles north-west of Nevada City. Elevation 2700'.
> Bibliography: Cal. State Min. Bur. Rept. XIII, page 254. U. S. Geol. Survey Folio 29, Nevada City Special.

The Mount Auburn patented claims comprise an area of 40 acres controlling 3000' along the lode. The Mount Auburn vein, which is the north continuation of the Merrifield-Spanish vein, occurs in the granodiorite, the strike is N. 35° W., and the dip is 45° NE. It has a width of from 2' to 4' of quartz, which carries free gold, pyrite and some galena. The vein can only be traced for 200' on the surface, but has been opened in the Ragon and Empire workings to the north and the Spanish mine to the south. The Mount Auburn vein has been developed to a depth of 600'. by an inclined shaft. Levels were run at 200', 300', 400' and 600' below the collar of the shaft. On the 200', and 300' level drifts were driven northwest on the vein, but the 400' level has been extended farther than either of these, having been run 900' northwest of the shaft. The 600' level has only been driven 200 feet. Practically no exploratory work has been done to the south. A considerable amount of ore is reported to have been stoped above the 400' level, but the amount and value is unknown. The nearest operating mine is the Champion which is situated about 1½ miles south.

**Muller & Walling.** (See North Wyoming.)

**Murchie Mine** (Lone Star). Owner, C. F. Humphrey, Nevada City. Under bond (1914) to H. W. Miller, 175 Fifth avenue, New York.

> Location: Nevada City Mining District, Secs. 8, 16 and 17; T. 16 N., R. 9 E., 2 miles east of Nevada City. Elevation 2760'.
> Bibliography: Cal. State Min. Bur. Rept. XIII, page 250. U. S. Geol. Survey Folio 29.

This property controls the mineral rights under 432 acres of patented placer and agricultural land.

There are four well-defined veins on the holdings, known as the Big Blue, Independence. Lone Star, and Alice Bell. Very little exploratory work has, however, been done on the two latter. The Big Blue (Murchie) and Independence veins have been developed by an inclined shaft on the former, 1150' in depth. Owing to the fact that no maps are available, no detailed description of the workings can be given. Levels have been run at depths of 400', 500', 600', 700', 900', 1000' and 1150' below the collar of the Murchie shaft. On the 600' level a drift was driven for a distance of 150′ east of the shaft. On the 700′ level the east drift was only extended 80′, while to the west of the shaft the vein was explored for a distance of 634 feet. On the 900′ level 877′ and on the 1000′ level 876 feet of drifting was done. The drain tunnel was being cleaned out in June, 1914, preparatory to development work.

The veins occur in granodiorite with hard and well-defined walls, so little timbering is required. The Big Blue vein varies in width from 1′ to 10′ with an average of 4′. The strike is nearly east and the dip 85° N. The Independence vein, which outcrops about 800′ east of the shaft, strikes north and dips 36° W.; it probably intersects the Big Blue vein.

The ore consists of quartz, carrying free gold, pyrite, galena, sphalerite and tellurides of gold and silver. The sulphide content is about 2% and the concentrates are worth from $100 to $200 per ton. This ore carries a larger amount of silver than is usual in the Nevada City and Grass Valley districts. As ·is typical of the district, although ore-shoots in the mine are irregular, the ore is of high grade. Two main ore-shoots have been developed, one 350′ east and the other west of the shaft on the 700′ level. They have both been stoped above the 1000′ level. The east ore-shoot was 500′ in length, averaged 6′ in width and the ore is said to have yielded from $15 to $20 per ton. From 1878 until 1884, 38,000 tons of ore were milled, yielding $587,000. The 700′ west stope is reported to have produced 6200 tons which averaged over $20 per ton. Ore mined between 1902 and 1910 was said to average $20 a ton.

The Murchie was a comparatively small dividend-paying unit at the time of its closure, which was caused by the operators becoming embarrassed financially through their widespread mining operations.

The total production of the mine to date is said to have been in the neighborhood of $1,150,000.

(Unwatering the mine began in May, 1914, but by August of that year had reached only the 400′ level, at an expense of $20,000. Work was later abandoned and the property was idle in September, 1918.)

**National Mine** (Constitution, Jesse Consolidated, Mammoth). Owner, National Gold Mining Company; Jesse R. Tautphaus, 25 Delmar street, San Francisco.

Location: Nevada City District, Secs. 9 and 10, T. 16 N., R. 9 E., 2½ miles east of Nevada City. Elevation 3000′.
Bibliography: Cal. State Min. Bur. Rept. XIII, page 249. U. S. Geol. Survey Folio 29, Nevada City Special.

The National mine, which consists of three claims, the Great Eastern, Great Western and Constitution, has been held by the present owner for many years, but very little development work has been done. On the Constitution claim, there are a number of small parallel fissure veins which occur in the siliceous slates. The vein which has been worked has an east strike; dips to the north and is from 1′ to 2′ in width. The ore, which is similar to the adjoining Lecompton and Federal Loan mines, contains abundant sulphides consisting of pyrite and arseno-pyrite, with smaller amounts of chalcopyrite, galena and zincblende.

**National Group.** Owner, Mrs. A. B. McKillican, San Leandro. Mine under bond to Wm. McLean, Graniteville.

Location: Graniteville Mining District, Secs. 8 and 17, T. 18 N., R. 11 E., 2 miles southwest of Graniteville. Elevation 4500′.
Bibliography: Cal. State Min. Bur. Repts. XI, page 309; XII, page 194; XIII, page 255. U. S. Geol. Survey Folio 66, Colfax.

The National property consists of two claims: The National, patented, and the National Extension location. These comprise an area of about 40 acres and a length of 3000′ along the Republic-National lode. The vein was first prospected on the National claim about 1875, but most of the development work was done between the years 1890 and 1898. During this time, the main tunnel was driven south on the vein a distance of 900′, where it connected with a 680′ tunnel which had been driven from the south side of the ridge by the owners of the adjoining Republic claims. The vein was thus prospected from one side of the ridge to the other for a distance of 1600′ and to a maximum depth of 368′ below the surface.

A shaft was sunk to a depth of 180′ and levels were driven south on the vein at a depth of 100′ and 170′ below the collar. At the 170′ level the shaft is said to have been in the foot-wall of the ledge and the drift was driven 300′ in a southerly direction before encountering the vein and pay shoot. The ore-shoot on this level was 160′ in length and from 3′ to 12′ in width. The vein strikes N. 3° E. and dips 78° E. Both walls are black slate, and there is a heavy gouge on the foot-wall. The ore is quartz carrying free gold with 1% to 2% sulphides. In the pay shoot, which was stoped from the bottom level to the surface, a distance of 320′, it is said to have had an average value of from $6 to $8 per ton in free gold. Ore of this value can be worked at a handsome profit in this district, since the cost of mining and milling should not exceed $2.50 per ton in normal times owing to the exceptional low cost of water power and timber and the abrupt topography which permits the veins to be worked by means of adits. Under certain conditions $1.50 and $2.00 ore has been worked at a profit in this region.

On the southerly adjoining Republic claim, which is also being operated by Wm. McLean of Graniteville, an adit has been driven 12,000′ on the vein, which at this horizon dips 45° E. This adit 650′ below the outcrop has already entered the National claims and it is expected to encounter the south raking National ore-shoot within a few hundred feet, giving a depth of approximately 300' below the lowest workings in the National mine. There is no equipment on the National property with the exception of an old 5-stamp mill. For further description of the Republic-National Mines, see Republic. It was reported in September, 1918, that McLean was installing a tramway from tunnel to mill.

**Nevada City Mine.** (See Champion.) Owner, North Star Mines Company.

Location: Nevada City Mining District, Sec. 11, T. 16 N., R. 8 E., 1 mile west of Nevada City.
References: Cal. State Min. Bur. Repts. VI, pt. II, page 49; VIII, page 418; XI, page 288; XII, page 190; XIII, page 245. U. S. Geol. Survey Folio 29, Nevada City Special.

**Nevada County Claim.** Owner, G. G. Allen Estate, Nevada City.

Location: Nevada City Mining District, Sec. 12, T. 16 N., R. 8 E., located in Nevada City townsite. Elevation 2500'.
Bibliography: Cal. State Min. Bur. Register of Nevada County Mines. U. S. Geol. Survey Folio 29, Nevada City Special.

The Nevada County vein was first worked in 1866 by means of a shaft 230' in depth, situated on the south side of Deer Creek, near the suspension bridge. Drifts were driven on the vein at depths of 100' and 200' for maximum distances of 700' north and 500' south of the shaft. The vein, which lies 1500' east of and parallel to the Reward-California Gold Tunnel vein, strikes N. 40° E., dips 50° E. and varies in width from a few inches to 2 feet. The ore from the main shoot near the shaft is said to have had a value of from $20 to $40 per ton. No work has been done on this vein for many years.

(The mine was abandoned long since. The stoping was done under town lots and the ownership and rights involved were obscure.)

**Never Sweat Mine.** Owner, Joseph Fischer, Nevada City.

Location: Nevada City Mining District, Sec. 9, T. 16 N., R. 9 E., 3 miles east of Nevada City. Elevation 3100'.
Bibliography: Cal. State Min. Bur. Rept. XIII, page 256. U. S. Geol. Survey Folio 29, Nevada City Special.

There are two claims comprising this property—the St. Louis and West Extension St. Louis. The Never Sweat vein has been developed by a 1600' tunnel, driven on the vein, which connects near its end with a 300' inclined shaft. The vein occurs in granodiorite which forms both walls; it varies in width from 3 inches to 2 feet. Its strike is N. 73° E. and, to a depth of 200' the dip is 45° N. At that horizon a fissure dipping to the south was encountered, and the vein is said to have left the original north-dipping fissure and followed this south-dipping fissure. The ore-shoot is two hundred feet in length and it has been stoped from the tunnel to the surface. Some rich ore is said to have been taken out. Besides the usual sulphides, antimony is said to have been present in the ore, which also carried a large amount of silver. The mine was idle in 1914.

**New Idea.** (See Syndicate Consolidated.)

**New Ophir Claim.** Owner, North Star Mines Company, 22 Williams street, New York City, and Grass Valley.

Location: Grass Valley Mining District, Sec. 1, T. 15 N., R. 8 E., 2 miles south of Grass Valley. Elevation 2700'.
Bibliography: U. S. Geol. Survey Folio 29, Grass Valley Special.

The New Ophir patented claim lies south of the Fillmore and Basin claims of the Sultana group. The Rich Hill vein of the Empire mine, which adjoins the Fillmore claims on the north, can be traced from the Empire claims through the entire length of the Fillmore, Basin and New Ophir claims.

On the New Ophir the strike of the vein is north and the dip 30° W. It has been prospected at intervals by shallow shafts and open cuts, but the greatest depth attained below the outcrop was only 45 feet. Both walls of the vein are diabase. No work has been done on the claim since 1905, and the superficial exploratory work prior to that time had failed to develop any outcropping ore-shoots. The ore-shoots in the district may occur at any horizon and on account of the potential value of this vein in depth, the property is worthy of more extensive exploration.

(It has recently been acquired by North Star Mines Company and was idle in September, 1918.)

**New Rocky Bar.** (See North Star.)

**New York.** (See Texas.)

**New York Hill.** (See North Star.)

**Niagara Mine.** Owner, Niagara Mining Company, 33 Canfield Building, Santa Barbara; Magnus Johnson, president; J. L. Hurlburt, secretary; L. L. Battey, manager.

Location: Rough and Ready Mining District, Sec. 18, T. 16 N., R. 8 E., 4½ miles west of Grass Valley. Elevation 1800' to 2500'.
Bibliography: U. S. Geol. Survey Folio 18, Smartsville.

The Niagara mine is located on the north side of Deer Creek, and consists of the Manila, Smith, Niagara Gold and Silver and Whitelaw-Reid claims, comprising an area of 60 acres. The vein can be traced at intervals for a distance of 800' on the surface. The walls of the vein are mainly granodiorite. It is reported that the area of amphibolite, which lies to the south of the granodiorite, was encountered in the foot-wall of the vein in the east levels. The general strike of the Niagara vein is east and the dip 30° to 45° N. Former operators sank an inclined shaft to a depth of 375 feet. The No. 2 level, 164' below the collar of the shaft, was driven east a distance of 248' and west 174 feet. A 105' raise was made at a point 60' east of the shaft; there is also a raise and a small stope 100' west of the shaft on this level. On the 300' level, 96' below No. 2 station, the east drift is 141' in length and at a point 61' east of the shaft there is a raise, the face of which is 35' above the level. The amount of gold taken from this vein is unknown, but the total is probably not large. There is an old 15-stamp water-driven mill on the property and a blacksmith shop. At the time the property was visited, June, 1914, a crosscut adit was being driven to cut the Manila vein upon which a small amount of surface work had been done. It was expected that this adit which was in 350' would encounter the vein within a short distance. The mine was not operated during 1915.

**Nichols Group.** Owners, H. B. Nichols and Pearl Voss, Grass Valley.

Location: Grass Valley Mining District, Sec. 28, T. 16 N., R. 8 E., 1½ miles west of Grass Valley. Elevation 2400'.
Bibliography: Cal. State Min. Bur. Register of Nevada County Mines. U. S. Geol. Survey Folio 18, Smartsville.

This property consists of the Nichols and Onion placer mines and the Nichols quartz claim, comprising in all an area of 80 acres of patented ground. The mine was not being worked in 1914.

**Norambagua Mine** (Three Sevens). Owner, Three Sevens Mining Company; Theodore R. Pell, president, New York City.

Location: Grass Valley Mining District, Secs. 11 and 15, T. 15 N., R. 8 E., M. D. M., 4 miles south of Grass Valley. Elevation 2100'.
Bibliography: Cal. State Min. Bur. Rept. XIII, page 256. U. S. Geol. Survey Folio 18, Smartsville.

The Norambagua mine is situated about 1¼ miles south of the Allison Ranch mine, near the southern extremity of the Grass Valley area of granodiorite. The property consists of the irregular-shaped Norambagua claims of 50 acres and the Norambagua No. 2 claim, 20 acres in extent. These claims control a length of 2500' along the lode. The mine was worked extensively during the 60's and is credited with a production of $1,000,000. After its closure about 1868, it remained idle until 1892. From 1892 to 1909 it was worked at intervals by the owners and tributers.

The vein in the early days was worked by a 567-foot inclined shaft but owing to the flat dip of the vein the vertical depth of the bottom of the shaft was only 120 feet. Levels were driven 500' north and 1000' south of the shaft. Development work in later years, however, has been carried on through a crosscut adit driven eastward from the bank of Wolf Creek. This adit encountered the vein at about 1200', and drifts were driven both north and south, the south drift connecting with the old workings.

From the adit level a shaft was sunk on the vein to a depth of 250 feet. The 100' level was driven 300' to the south and there is a small stope 50' from the shaft. On the 250' or bottom level, the south drift has been extended 700 feet. At a point 450′ south of the shaft a winze was sunk to a depth of 150′ and 'rich specimen ore' is said to have been encountered.

(In May, 1918, it was reported that the present company, which had been doing exploratory work for about two years, had found a promising vein at the 700-ft. and 800-ft. levels.)

The general strike of the vein is N. 30° E. but, unlike the Allison Ranch-Omaha veins occurring in the same area of granodiorite, and dipping to the west, the Norambagua vein dips to the east. Near the surface the dip was from 15° to 20°, but below the tunnel level the vein steepens, reaching a maximum of 35° to 40° E. Both walls are grano-diorite, with comparatively fresh rock within a few inches of the vein. The width varies from a seam to a maximum of 10 inches, the average being approximately 5 inches. The ore, which is said to have had a value of from $40 to $100 per ton, carries free gold finely disseminated, with pyrite and arsenopyrite which give the quartz a banded structure. No ore shoots of note have been encountered in the lower workings. The equipment consists of a blacksmith shop, small compressor, pump and hoist. The mine was unwatered for examination in 1915.

(It was idle in September, 1918, and was reported to have been closed for the period of the war.)

**Normandy Dulmaine Group.** Owner, Normandie-Dulmaine Gold Mines Company, Grass Valley, Nevada County; Frank Dulmaine, president; Horace Jones, manager.

Location: Rough and Ready Mining District, Sec. 5, T. 15 N., R. 8 E., and Secs. 31 and 32, T. 16 N., R. 8 E., 3½ miles southwest of Grass Valley. Elevation 2200′.
Bibliography: Cal. State Min. Bur. Rept. XIII, page 257. U. S. Geol. Survey Folio 18, Smartsville.

This property includes the Dulmaine, W. Dulmaine (fraction), Normandy, W. Normandy, E. Normandy and South Normandy claims, none of which are patented. They cover an area of about 70 acres and 3000′ along the Normandy and 1500′ of the Dulmaine vein. The sub-district in which this mine is located is locally known as Deadman's Flat and it was renowned in the '50's for the rich alluvial placer deposits. The gold of these placers was undoubtedly in part derived from the weathering and erosion of the numerous veins in the vicinity which have since become noted for their 'rich pockets' of coarse gold. The Normandy-Dulmaine mine was owned and operated by Hunt and Talbot from 1871 to 1886. Ore is said to have been taken from the mine during this period which yielded from $40 to $489 per ton. Later the Normandy was relocated by A. Seneclial and in 1896 it was pur-chased by Frank Dulmaine, who organized the Normandy-Dulmaine Gold Mines Company in 1912. The mine is credited with a total production to date of $100,000, but the actual amount produced is probably somewhat less than this. There are three well-defined veins known as the Dulmaine, Normandy and West Normandy. The Dulmaine vein strikes N. 25° W. and dips nearly vertical. The Normandy lies 700′ east and is parallel in strike to the Dulmaine. Near the center of the Normandy claim this vein divides into two branches which unite 500′ further north. At the surface the two branches are only fifty feet apart, but the dip of the western branch is 35° W., while that of the eastern or foot-wall vein is 65° W. Owing to the rapid divergence of these veins, at a vertical depth of 450′ below the surface they will be over 500′ apart and the west vein will have intersected the Dulmaine vertical vein, provided they continue on the same dip as shown at the surface.

The logical way to develop the veins would, therefore, be to sink a 400′ vertical shaft on the Dulmaine vein to its point of intersection with the west branch of the Normandy vein and then crosscut 500′ east to explore the foot-wall branch. These veins occur in a complex of gabbro, diorite, diabase and amphibolite, which is a derivative of the other rocks. The well-defined walls of the fissure are from 2′ to 4′ apart, but the quartz varies from a few inches to 2′ only in width, altered wall rock forming the rest of the fissure filling. The gold and auriferous sulphides occur in the quartz, and there is a considerable amount of heavy, coarse gold or 'specimen ore.' The sulphides consist of pyrite, chalcopyrite and galena. In addition to the veins mentioned, there is a 'cross vein' which intersects the Normandy and Dulmaine ledges near their southern ends. This vein, which is known as the West Normandy, strikes east and dips 80° S.

All of the veins have been worked at different points to shallow depths by open cuts and prospect shafts. The deepest workings consist of an 80′ vertical shaft from which, at a depth of 70′, a crosscut was driven east for a distance of 42′, where the foot-wall branch of the Normandy vein was encountered. Drifts were extended 135′ north and 182′ south, and the quartz is said to have averaged 18″ in width. At a point 145′ south of the crosscut very hard rock was encountered and the vein 'pinched,' but the last 10′ of the drift is in 'softer' rock and the vein is reported to be opening out again. According to Mr. Frank Dulmaine, an inclined shaft, located 100′ north of the vertical shaft, was sunk 110′ on the west vein in 1894; drifts were run on the vein 100′ north and 30′ south and $2800 in 'specimen ore' taken out.

The equipment on the property in 1914 was negligible and it will be necessary to install electric power and a new plant before exploratory work can be continued.

**North Banner Consolidated Mine.** Owner, North Banner Consolidated Tunnel Company, 525 Crocker Building, San Fra  i..:o; Chas. E. Green, president; I. N. Rosecrans, secretary.

Location: Nevada City Mining District, Sec. 16, T. 16 N., R. 9 E, 2 miles east of Nevada City. Elevation 3300'.
Bibliography: Cal. State Min. Bur. Repts. VIII, page 420; XII, par.  ''; XIII, page 257. U. S. Geol. Survey Folio 29, Banner Hill.

The patented quartz claims known as the Woodville, North Banner and Kohinoor, 55 acres in all, comprise the holdings of this company. The property lies on the northwestern slope of Banner Mountain, one of the most prominent features of the topography of the district. It has been held by the present owners since 1883. From 1889 to 1892 the mine produced $175,000; from 1892 to 1896 it was operated at intervals, the last bullion being produced in the latter year. In 1904 the mine was operated for a short period, but no ore was treated. Since that time it has been idle. There are four well-defined veins. The Woodville, which is the most westerly, has been developed by means of a 1000' adit, near the end of which a shaft was sunk on the vein to a depth of 500 feet. The greater part of the gold produced has been taken from the Woodville vein, as the others have been only superficially developed. This vein strikes N. 38° W., dips 30° to 45° NE. and averages about 2' in width. The Dunnington vein lies 100' east of the Woodville, the Tiny, 250' east of the Dunnington, and the Reindeer, 150' east of the Tiny. Like the Woodville, all dip from 35° to 45° E., but their strike is nearly north. All of the veins outcrop in a small area of diorite bounded on the south by the Calaveras slates and overlain to the north and west by a capping of andesite. There was one well-defined ore-shoot of irregular shape in the Woodville vein which was stoped from the bottom of the shaft to the surface. It is reported that the 500' or lowest level was driven 400' south in good ore. The ore is quartz, carrying free gold and 5% of sulphides, consisting of pyrite, galena, and smaller amounts of molybdenite and tetrahedrite. The sulphides were worth $160 per ton. The ore in the Woodville vein is characterized by its high percentage of silver, the silver equalling by weight the gold content. There is no equipment and the mine was idle in 1914.

**Northern Bell Group.** Owners, J. M. Thomas et al., Grass Valley, California.

Location: Grass Valley Mining District, Secs. 1 and 12, T. 15 N., R. 8 E., 2 miles south of Grass Valley. Elevation 2800'.
Bibliography: U. S. Geol. Survey Folio 29, Grass Valley Special.

This property, situated on the western slope of Osborne Hill, consists of the Northern Bell patented claim, 20 acres in extent, and the Thomas patented fraction of 10 acres.

The claims are located at the extreme southern end of the Osborne Hill fissure system. Very little work has been done. A shaft was sunk on a 2' vein to a depth of 130' and several short tunnels have been driven, but no pay shoots were encountered. The vein strikes north and dips from 25° to 45° W. The walls are hard diabase. Some 'good ore' is said to have been produced in early days. The nearest mines upon which extensive work has been done are the Conlin and Osborne Hill, lying one-half mile to the north.

**North Star Mine** (includes the following mines: Massachusetts Hill, Gold Hill, New York Hill, New Rocky Bar, Granite Hill, Cincinnati Hill and others). Owner, North Star Mines Company, 22 Williams street, New York; Geo. B. Agnew, president; W. D. Pagan, secretary; A. D. Foote, manager.

> Location: Grass Valley Mining District, Secs. 2 and 3, T. 15 N., R. 8 E., and Secs. 27, 28, 33 and 34, T. 16 N., R. 8 E. Mine adjoins townsite of Grass Valley. Elevation 2400'.
> Bibliography: Cal. State Min. Bur. Repts. VI, pt. II, page 44; VIII, page 428; X, page 376; XI, page 270; XII, page 195; XIII, page 257. U. S. Geol. Survey Folio 29, Grass Valley Special.

The immense holdings of the North Star Mines Company (covering an area 3 miles long and from 1 to 2 miles in width and containing over 1200 acres) lie west and southwest of Grass Valley. They have been accumulated around the famous North Star mine; or, as someone has so aptly put it, "the North Star has swallowed" the adjacent mines and mineral rights. Among the properties which have been absorbed by the North Star Mines Company, are, in their order of prominence: the Massachusetts Hill, Gold Hill, New York Hill, New Rocky Bar, Granite Hill, Larimer and Cincinnati Hill. Mr. Wm. Hague has estimated "that the mines now owned by the North Star Mines Company (excluding the Champion group of mines in the Nevada City District) have produced from 1850 to 1913 inclusive, a total of from $25,000,000 to $27,000,000, of which amount the North Star vein has yielded $16,715,000 from 1,170,000 tons of ore. There are only two properties in the district which have equalled or surpassed this production; one is the Empire mine, which has been worked continuously since 1850, and the other is the famous Eureka-Idaho vein, the authentic production of which has been over $19,000,000 from one ore-shoot.

The history of the North Star is interesting and instructive. The North Star vein was discovered by the Lavance brothers in the fall of 1851. A number of claims one hundred feet square were located by different men and these, a few years later, were consolidated as the Helvetia and Lafayette Gold Mining Company. This group of claims controlled 480' along the apex of the vein. Adjoining on the east was the Independence group, covering 800' of the vein. The White Rock property 800' in length was located east and adjoining the Independence. In the hanging wall of the Helvetia and Lafayette mine a group of claims had been located known as the North Star. In 1858 the Helvetia and Lafayette mine was sold at sheriff sale for $8,000 to Edward McLaughlin who, two years later, sold the property for $15,000 to Edward Coleman et al. who were the owners of the North Star property. In 1861 the North Star Quartz Mining Company was organized and during the next five years many adjoining claims were purchased. According to estimates of Mr. Hague the total production of the North Star vein from 1851 to 1866 was $1,500,000; the ore during this period having an average value of $25 per ton. Of this sum probably $1,000,000 was produced during the five years from 1861–1866. In 1867 the holdings were consolidated by the North Star Gold Mining Company. The North Star claim, which is 3000' in length and 800' in width, was patented in August, 1875. From 1867 to 1875 the property is credited with a production of $1,125,000 from 50,000 tons of ore. By 1875 the North Star inclined shaft had reached a depth of 1200', at which depth the ore decreased in value until the mine could not be worked at a profit. Operations were therefore suspended in 1875 and the property remained idle until May, 1884. In the latter year the property was reopened by W. B. Bourn, who had formed a company known as the North Star Mining Company. A few years previous to this Mr. Bourn had been successful in putting the Empire mine on a paying basis after it had been condemned as 'worked out' by a number of 'mining experts.' Nevada County and especially the Grass Valley district are greatly indebted to Mr. Bourn for the rejuvenation of mining at Grass Valley when practically everyone else thought that the mines of the district were exhausted. In May, 1884, the North Star shaft was reopened and within a short time the mine was again on a paying basis. In the following ten years the North Star inclined shaft was sunk from the 1200' to the 2400' level. The ore-shoot, which was supposed to have been worked out, was again encountered below the 'barren zone' and some of the richest ore the mine has produced was taken from the 1800' level near a 'crossing' or fault; this ore-body continued to below the 2100' level and 250,000 tons of ore yielded $5,250,000 and in 1893 the average yield was $31 per ton. Below the 2100' level the vein 'pinched' to practically a seam. Although at this period there was no ore in the lower levels of the North Star mines, James D. Hague showed his faith in the district and in the recurrence of payshoots in depth by advising the sinking of a 1600' vertical shaft in the hanging wall of the North Star vein. This shaft, known as the Central, was started in 1897 and the vein was encountered March 31, 1901, at a vertical depth of 1630' or an inclined depth of 4000' below the collar of the North Star incline. From the bottom of this shaft, drifts were driven east and west and a main raise was started, which connected with the bottom of the North Star inclined shaft in November, 1903. In the meantime, from 1894 to 1900, a number of lean years had been experienced by the North Star mine, but during this period the Massachusetts Hill mine was reopened by the North Star Mines Company and produced $1,000,000 from 68,000 tons of ore. In 1900 the 500' level of the North Star mine was driven eastward and a small but very rich ore shoot was discovered 1500' east of the shaft.

The top of this pay shoot proved to be only a short distance above the 500' level. From this point upward the vein narrowed down to a seam, and there were no surface indications to warrant an assumption that such a rich shoot existed at a depth of 500 feet. Although this ore-shoot did not extend above the 500' level, it was successively opened up in depth by the 1100', 1900' and 2700' east drifts. From the year 1900 to 1911, 182,000 tons of ore were stoped from this pay shoot, 18,000 tons of which yielded an average of $37.50 per ton. From 1901 to 1906 the development work prosecuted through the vertical shaft was confined to driving drifts east and west from the main raise on the 4000', 3900', 3700', 3400' and 3000' levels. In 1906 sinking was begun on the vein from the bottom of the vertical shaft and in 1915 this incline had reached a depth of 2300' below the 4000' level. The North Star vein had thus been developed to an inclined depth of 6300' or to a vertical depth of approximately 2500 feet. Levels have been driven east and west of the shaft from the 4400', 4700', 5000', 5300', 5600', 6000' and 6300' stations. From 1901 to the end of 1912, 43,192' of development work was done below the 2700' level, and 635,028 tons of ore were hoisted therefrom. The following statistics are quoted from Mr. Hague's description of the North Star mine:

"Of the 635,028 tons hoisted from the deeper levels, 626,509 tons were crushed, of a gross value of $7,496,499.73, compared to an estimated yield from the surface to, and including, the 2700-foot level of 438,273 tons of a gross value of $8,019,016, making a total production of the North Star mines to the end of 1912 of 1,064,782 tons, of a gross value of $15,515,515. The enterprise as now conducted has been practically continuous since 1884, and precise figures can therefore be given for a period of over 28 years. The following comparison may be made between the results of the operations of the upper and lower levels of the mine:

Upper mine operations to and including 2700-foot level (1200 feet vertical), from 1851 to end of 1913 (operations of the upper levels ceased in 1911):

Production (438,000 tons at $18.308) _____ $8,019,000 00
Operating and development expenses (per ton $12.90) _____ 5,650,000 00

Profit (per ton $5.408) _____ $2,369,000 00

Vertical shaft operations, 3000 to 5300-foot level (2000 feet vertical), from 1901 to end of 1913:

Production (732,599 tons at $11.871) _____ $8,696,596 15
Operating and development expenses (per ton $6.383) _____ 4,676,371 43

Profit (per ton $5.488) _____ $4,020,224 72

Total North Star vein (surface to 5300-foot level), from 1851 to end of 1913:

Production (1,170,000 tons at $14.286) _____ $16,715,000 00
Operating and development expenses (per ton $8.825) _____ 10,326,000 00

$6,389,000 00

## PRODUCTION IN DEPTH.

"The tables _____ show a decline in yield per ton with increased depth, but to offset this it appears that the yield per square foot (horizontal) mined has increased. This may be explained by the fact that in the upper portions of the mine above the 2700-foot level both the vein and quartz were narrower, with fewer stringers than below, although definite figures on this point can not now be given. To lose now as few stringers as possible, much waste, is unavoidably sent to the mill, thereby reducing the yield per ton from what it would be if only clean quartz were treated. This course seems to be justified by the result, as a greater total yield for the vein as a whole is made available. The North Star vein is now exhausted, so far as is known, above the 3000-foot level. The total area mined above this level is in round numbers 2,100,000 square feet of a total of 6,000,000 square feet prospected, or, in other words, 35% of the exploited ground has proved profitable. In the report made by Clayton in 1869, he stated that with good management there was no reason why the cost, at this time, of $16 to $20 per ton, should not be reduced to $12 cost per ton. For the five years ending with 1892 the cost, including improvements, was $12.35 per ton. Since then the cost has been dropping until now it is about $5.25 per ton mined. The profit per ton has remained nearly constant since 1884."

Photo No. 29. Hoist at Central Shaft, North Star Mines. Photo by C. A. Waring.

## Geology.

The North Star and associated veins form a system of east-striking, north-dipping veins as distinct from the Empire-Pennsylvania-Osborne Hill and Omaha-Allison Ranch systems of north-striking, west-dipping fissures and veins. The origin, age and relation of the various fissure or vein systems of the Grass Valley district are more or less obscure. All the data so far disclosed, however, seems to lead to the conclusion that the north-south west-dipping fissures, of which the Empire and Pennsylvania veins are examples, are the oldest; that the North Star group are of a slightly later period; but that both systems owe their origin to the same dynamic compressive stress applied from different directions.

The North Star vein, together with the parallel and divergent veins lying both in the foot and hanging wall of the 'main' North Star vein,

Photo No. 30. The North Star Vein, showing—A. Hanging wall of hard diabase; B. Gouge; C. Quartz (main vein); D. Horse of altered diabase, traversed by stringers of quartz and calcite; E. Footwall vein; F. Diabase footwall.

outcrop in the western area of diabase-porphyrite. (See description of Grass Valley district.) As is characteristic of most of the veins in this district, the outcrops of the veins are inconspicuous and can only be traced at intervals on the surface. At a depth of from 3700 to 4000' on the dip, the North Star vein passes from diabase into the central core of granodiorite, in which also occur the Pennsylvania W. Y. O. D., Omaha-Allison Ranch veins and the Empire vein below the 1700' level. As in the other mines of the district, the different character of the wall rocks does not appear to have any effect upon the size of the vein or upon the mineralogical character or tenor of the ore. The walls of the North Star fissure within the ore shoots are as a rule from 2' to 4' apart, but outside of the payshoots the vein may narrow for hundreds of feet to a mere seam. As a rule, the space between the walls is filled with crushed and altered wall rock and auriferous quartz; generally the quartz does not occupy more than from 1' to 3' of this space. The fissure and vein vary in form from the usual undulating simple fracture to a complex system of interlinked veins. In some cases the vein will 'split' and these branch veins may or may not reunite with the main fissures. In a number of cases two orebodies have been formed separated by 'horses,' or false walls of country rock. The typical vein structure of the North Star is shown in the accompanying photograph. The complicated structure of the fissures in the Grass Valley district has been responsible for the closing of many properties. The secret of successful mine operations is to do an apparently excessive amount of exploratory work so that no branch parallel vein or orebody may be overlooked. While the North Star Mine is supposed to be exhausted above the 3000' level, it is within the realms of possibility that intelligently directed exploratory crosscuts into the foot and hanging walls may develop other veins or orebodies; especially is this true of the territory west of the North Star shaft, where the vein and ore shoot practically terminate along a 'crossing' or fault plane which strikes north and dips 50° W.

In the Grass Valley-Nevada City district the ore shoots are irregular in size and shape and may occur at any horizon from the surface to the deepest levels yet attained. Orebodies 'apex' at different levels and occur at irregular intervals in the 3000' explored along the strike of the lode. There is apparently no connection between or systematic arrangement of, the various pay shoots; and they may be separated both horizontally and vertically by hundreds of feet of barren ground throughout which the fissures may close down to a mere seam. These so-called 'barren zones' are also encountered in the other mines of the district, notably in the Empire mine, where practically no ore was developed between the 1500' and 2100' levels. It has, therefore, been necessary in the North Star and other mines to develop immense areas of unproductive territory in order to find payshoots. However, when one profitable orebody has been found in a vein system, it is almost an assured fact that systematic development along the plane of the vein both horizontally and vertically will sooner or later develop other ore shoots and although hundreds of feet of costly drifts and crosscuts may have to be driven through barren ground, when one of the rich shoots is encountered, the profit derived therefrom makes the development cost negligible. An idea of the rich- ness of the ore in these shoots may be gained from the fact that, from an area of only a few square feet, 'specimen' to the value of $60,000 was taken out, in twenty candle boxes, from the 3700' level which has so far been one of the most productive levels in the mine.

**The ore.**

As has been previously stated, the vein filling is composed of altered country rock and quartz carrying auriferous sulphides, consisting for the most part of pyrite with small amounts of galena, sphalerite and chalcopyrite. Finely disseminated galena and sphalerite are generally found associated with the rich 'specimen ore.' The gold occurs in some cases as coarse leaves and seams in the massive quartz or it may be disseminated throughout the quartz in fine particles associated with the sulphides. In the upper levels the brecciated and altered diabase enclosed between the walls contains a large amount of carbonates. The accompanying photograph (No. 30) shows stringers of calcite thread- ing a horse of diabase which is included between the foot-wall and hanging wall veins. The auriferous quartz may occur either massive or banded by subsequent movement and successive deposition of the sul- phides. There seems to be little relation between tenor of the ore and the depth at which it is found. In the weathered or oxidized zone, which extended only to a depth of from 5' to 150' very high grade ore was taken running from $50 to $100 per ton; below this zone, however, the ore averaged from the surface to the 3000' level in the neighborhood of $20 per ton; below the 3000' level it averaged $11.871, owing to the fact that more barren wall rock or 'waste' was included in the ore from the lower levels. 'Specimen ore' which runs thousands of dollars per ton is not confined to any horizon, but has been found in the deepest levels yet attained. The sulphide content of the ore is about 2% and the concentrates have a value of from $60 to $125 per ton.

The holdings of the North Star Mines Company include a number of now dormant mines, among which are the Massachusetts Hill, Gold Hill, New York Hill and Granite Hill, all of which have been prolific producers in the past and have merit.

The Gold Hill vein was discovered and located by George Knight in September, 1850. This is supposed to have been the first discovery of auriferous quartz in Nevada County. On the top of Gold Hill a large amount of rich quartz was found; the surface had been enriched by concentration of the gold by weathering and erosion from the stringers and veins. It is commonly reported that over $2,000,000 was taken from the shallow surface workings. The Gold Hill mine is credited with a total production of $4,000,000 from 1850 to 1867. The mine was also operated for a few years following 1890. The Gold Hill property was purchased in April, 1903, by the North Star Mines Company, and from that time until 1906 the vein was opened through the inclined shaft to an inclined depth of 1200 feet. A large amount of lateral development work was done, but only a few scattered pay shoots were encountered and the mine was finally closed.

The Massachusetts Hill vein has the same general north strike and the same dip to the east at an angle of 25° to 30° as the Gold Hill vein, of which it is the southern continuation. It was discovered in 1850 and worked with but few interruptions until 1866. In 1894 the property was purchased by the North Star Mines Company. The production prior to the year mentioned is estimated to have been from $4,000,000 to $5,000,000 from 125,000 tons of ore.

The orebody from which this gold was taken extended from the surface to a depth of 600' where the usual 'barren zone' was encountered. In 1896 the mine was reopened and a new ore shoot was developed at a depth of 1200', which yielded 68,222 tons of ore from which $1,078,075 was recovered. The shaft was extended to the 1800' level, but exploratory work at this point failed to find ore and the mine was closed in 1901. Since that time no work has been done on this vein.

The New York Hill vein belongs to the North Star system of fissures which strike east and dip north, and lies from 1500' to 2500' in the hanging wall of the North Star vein. The strike of the vein is irregular, but the general direction is west and northwest. Its width is from 8" to 2' included between hard walls of diabase. The New York Hill was one of the pioneer mines and is credited with a production of $500,000 between 1852 and 1865. In the following two years 2189 tons of ore yielded $106,430. The mine was closed about 1867, but was reopened in 1874, and $100,000 was taken out from September, 1874, to October, 1875, the ore varying in value from $28 to $49 per ton. At the time the mine was closed in 1883 the shaft had reached a depth of 1300' on the vein. The North Star purchased the property in 1894, subsequent exploratory work failed to develop new ore shoots and after $61,263 had been taken out the mine was closed. Northwest of the New York Hill occurs the 'Hogsback' or New Rocky Bar vein. The underground workings on this vein show the intimate relation which exists between the conjugate vein systems and proves conclusively their contemporaneous origin. The New Rocky Bar vein does not outcrop, the highest point of the 'Hogsback' or anticlinal axis being about 100'

Photo No. 31. Steel headframe, 100 feet high to the sheaves; ore bins and sorting plant at Central Shaft, North Star Mine. Photo by C. A. Logan.

below the surface. The flat top of the anticline is 300' across and from it the northern limb dips 15° N., while the southern limb dips 25° S. The north-dipping vein has been developed by footwall crosscuts from the different levels of the Chevanne shaft on the New York Hill vein. The New Rocky Bar inclined shaft and the three drifts therefrom have explored the south vein to an inclined depth of 400 feet. For map and sections showing the complex relationship and development work, see "The Gold Quartz Veins of the Nevada City and Grass Valley Mining Districts," U. S. Geol. Survey 17th Annual Report, pages 237–238.

From 1880 to 1885 large amounts of 'specimen rock' containing much coarse gold were produced. It is estimated that $500,000 has been taken from these veins in a total of 35,000 tons. No work has been done since that time.

The Cincinnati Hill mine which was worked to a depth of 150' in early days, was purchased by the North Star Mines Company in 1910. After reopening and sinking the shaft to a depth of 400', 994' of drifts were driven on a 'large barren vein.' In 1912, 656 tons of ore were stoped, the yield from which was only $3,652. The exploratory work done, failed to encounter any ore shoots and the mine was closed in 1912.

The North Star quarterly report. issued in the early part of 1916, showed 24,414 tons of ore hoisted, as against 25,325 tons for the succeeding quarter. Miner's shifts were increased from 4133 to 5081. Efficiency of operation is being improved by hand-sorting of ore. thus decreasing the amount of rock going through the mill, and increasing the yield per ton.

In April, 1916. extensive developments were being prosecuted on the 6300' level, where a new orebody was encountered the previous summer. The vein is proving one of the best in the property and is expected to yield a large tonnage of excellent ore in the near future. Large quantities of ore are going to the two 40-stamp mills from the 3400', 4000', and 4400' levels. On the 3400' level a strong shoot of fair grade ore has been opened.

### Changes and additions to plant.

[The mine was visited in September, 1918, and much new data obtained. All work is now carried on through the Central shaft and all ore is milled and cyanided at the Central Mill and cyanide plant. The new mill, which replaces two formerly in operation, contains 60 stamps, 1500 lbs. each, and has a capacity of 300 tons daily. The consolidated milling is said to save 20¢ a ton. The cyanide plant handles the entire mill tonnage.

The steel head-frame. 100 feet high to the sheaves, weighs 96,000 lbs. and the steel bins, weighing 188,000 lbs. have a capacity of 500 tons of ore, and 150 tons of waste. The sorting plant is attached to the bins. The hoist is a Nordberg duplex air-engine, with Nordberg cut-off. The drums are 9 feet in diameter and have 6-ft. face, and the rope speed is 1200 ft. a minute. The load is 4 tons of ore in addition to the 5000-lb.

skip, and the depth capacity of the hoist is 7300 feet on the incline. The main compressor (No. 1) is a Laidlaw-Dunn-Gordon, direct connected to Pelton waterwheel 18½ ft. in diameter. It has variable-volume gear and compresses 2310 cu. ft. of free air per minute. In the power house there are also two compressors, each with capacity of 1300 cu. ft. free air a minute and one with capacity of 100 cu. ft. free air a minute.

### Mining.

A. B. Foote stated that up to January, 1918, no stoping had been done below the 4400-ft. level but that preparations were now being made to stope from all the deeper levels as far as 6300-ft. Mining was also being carried on in September, 1918, on levels above the 4000-ft., off the inclined shaft, and southeast of the vertical shaft off the 4000-ft. level. The contact of granodiorite and diabase was noted near the shaft at the 1600-ft. vertical (4000-ft. incline) level. The workings above here are in diabase which is extremely hard, but stands well. Stulls which have evidently been standing about 10 years in drifts on the 3400' and nearby levels show little or no sign of crushing, although some are entirely rotten. Many drifts have required no timbers for considerable distances. Many stopes have been carried up 300 ft. to 400 ft. on the dip. 'Go-devils' are largely used. The dip of the vein is so flat that rock will not slide, and has to be handled over and over. Splitting of the vein and occasional rolls add greatly to the difficulty of shoveling, by giving a rough, uneven floor in the stope. The vein pinches and swells, varying in width, from a streak of gouge to 3 feet, but averaging in the workings visited by the writer, 5" to 8". It is stated that the average gold content per ton does not fluctuate with the width of the quartz; six inches of quartz and three feet of wall rock may give the same return as two feet of quartz, and two feet of wall rock. The tendency is to carry a number of small stopes rather than a few large ones, as this makes for better working conditions.

The management states that approximately 35% of the ground explored on the North Star vein has proved profitable, requiring approximately one foot of development work to make available 20 to 25 tons of ore.

The vein shows less tendency to split in the granodiorite than in the diabase. The former rock is easier to work than the diabase and requires little timbering as a rule; but it occasionally happens that lines of weakness occur where the basic constituents have segregated out in bunches, giving heavy ground which tends to break away.

Stope bosses have charge of work and supplies, and handle the powder. Miners are paid $4 plus a bonus for all holes drilled after the fifth. Muckers are paid $3.50 plus a bonus of 15¢ a car after the sixth.

It was found by tests that close sorting underground was impossible. The smallest size that could be accurately sorted there was 6 inches in diameter. The sorting plant employs eight men and a foreman. The ore is washed as it goes onto the sorting table and very close work is possible, it being stated that the rejected rock contains far less than it would cost to treat it. Of 400 tons hoisted daily, 100 tons are sorted out. This results in sending a much higher grade product to the mill and adds several hundred dollars a day to output. In 1917 the average daily run of the sorting plant was only 5 hours and 8 minutes, and the men sorted out an average of 11 tons each of waste per shift. In 1918 the average run to September was 6½ hours daily and each man averaged 12½ tons a shift. The men are paid $3.25 a shift, and receive a bonus of 15¢ a ton for each ton sorted over 8½. With an occasional hour overtime they make an average bonus of 75¢ a day.

The operations in July illustrate the saving effected by sorting. That month 11,735 tons were mined and hoisted. From this 2993 tons were sorted as waste, at a cost of $1295, equal to 43¢ per ton of waste, or 11¢ per ton mined. Milling and cyaniding cost about $1.15 per ton mined during that month, and the rejected rock carried probably little over one-third this much per ton.

### Milling and cyanidation.

Ore is dumped from the skips over 1½″ grizzlies. The coarse from the grizzlies goes to a separate bin and is drawn onto the two balanced shaker tables, where it is washed and sorted, going thence to two Wheeling crushers. It is hauled to the mill in 2½-ton cars by an electric locomotive. The mill bins are of concrete and have a capacity of 1000 tons. Each ten-stamp unit has its own 35-horsepower back-geared motor. The rock is crushed to about 20-mesh and the pulp flows over 3′ by 16′ plates thence to Richards pulsator classifiers and to three 5-ft. cones. Slimes are piped to the cyanide plant. Sands are concentrated on seven Deister double-deck Simplex sand tables and middlings from these on Deister single-deck tables. Concentrate is ground in a pebble mill and run over an amalgamating plate, and tailing and concentrate then discharge into pipe lines and go to the cyanide plant. The slime is thickened in a Dorr thickener and treated in Pachuca agitators. The concentrate is thickened and treated in Devereaux agitators. Tailing is leached and mixed with slimes. Recovery is made in Merrill presses in which building paper is used in place of cloth, at a saving in cost.

### Disposal of tailing.

The tailing has in the past been discharged into the creek but the company was recently required to stop this. The cyanide plant is so near the creek that there is no storage room between. A large clam-shell bucket excavator was installed this summer (1918) at a cost of about $22,000 (see photographs). One terminal is a mast erected near the cyanide plant and the other is a movable tower on rails on the hill across the creek. New storage space can be obtained by moving the tower. The excavator is operated electrically by one man stationed in the tower, and tailing is carried to the slope on the east side of Wolf Creek. Estimating the cost of operation at 1½¢ a ton, and allowing for amortization in 10 years, it is thought that this method of disposal will add five cents a ton to working costs.

The following comparative figures for costs of certain mining items in 1913 and 1917 indicate strikingly the problem being faced by all operating quartz mining companies. Figures are for tonnage mined:

Mining Cost (Operating).

|  | 1913 | 1917 |
|---|---|---|
| Breaking rock | $0.875 | $1.12 |
| Timbering | .283 | .44 |
| Shoveling | .881 | .63 |
| Tools | .159 | .22 |
| Machine drills | .323 | .30 |
| Blacksmiths and mechanics | .108 | .04 |
| Lowering main raise | .130 | .12 |
| Repairing | .024 | .01 |
| Tramming | .173 | .18 |
| Hoisting | .209 | .18 |
| Pumping | .204 | .19 |
| Office, foremen, bosses | .217 | .24 |
| Miscellaneous | .139 | .13 |
| Totals | $3.180 | $3.80 |

Figures for total costs of operations in 1914 and 1917 were stated as follows:

|  | 1914 | 1917 |
|---|---|---|
| Mining | $3.043 | $4.629 |
| Milling | .621 | .611 |
| Cyaniding | .415 | .458 |
| Bullion | .030 | .034 |
| Miscellaneous | .299 | .364 |
| New York office | .190 | .158 |
| Taxes | .250 | .272 |
| Accidents | .093 | .145 |
| Tailing disposal |  | .028 |
| Total per ton treated | $4.941 | $6.699 |

A. B. Foote estimated in September, 1918, that total cost for the current year would be about $7.20 per ton, the principal increase being in mining costs.—*C. A. Logan.*]

Photo No. 32. System for digging and storing tailing at North Star Mine. The excavator is electrically operated by one man stationed in the tower in the background. The tower is on rails to permit easy moving. Photo by C. A. Logan.

**North Wyoming Mine** (Muller & Walling). Owner, North Star Mines Company.

Location: Nevada City Mining District, Sec. 2, T. 16 N., R. 8 E., 1¼ miles west of Nevada City. Elevation 2600'.
Bibliography: Cal. State Min. Bur. Repts. XI, page 296; XII, page 194; XIII, page 255. U. S. Geol. Survey Folio 29, Nevada City Special.

The property consists of the North Wyoming patented claim and two fractions, comprising an area of about 22 acres and covering 2300' along the strike of the lode. The vein, which is the Northern branch of the Ural-Nevada City vein in the granodiorite, has been developed only superficially in the North Wyoming holdings. There is an 85′ inclined shaft from the bottom of which a 50′ drift has been driven on the vein, and a 300′ crosscut tunnel has also been driven. The vein dips to the east and ranges in width from 2′ to 12′ with an average of 3 feet. The ore is said to average $10 per ton. The last work done was in 1904.

Property purchased early in 1915 by North Star Mines Company.

**Oakman Group.** Owner, Oakman Consolidated Mining Company; care of Frank Dillon, Nevada City, California; J. W. Watson, Moores Corner, Massachusetts; A. J. Arnstein, Shelbourne Falls, Massachusetts.

> Location: Washington Mining District, Sec. 31, T. 18 N., R. 11 E., 2 miles north of Washington. Elevation 3000′.
> Bibliography: U. S. Geol. Survey Folio 66, Colfax.

This group of 16 unpatented mining claims, 320 acres in all, lies near the junction of the two forks of Poorman Creek, joined on the north by the Spanish mine, and with the well-known Gaston mine about 2 miles east of the holdings.

The Lady Oakman vein upon which most of the development work has been done averages from 4′ to 8′ in width, dips 65° E. and strikes N. 10° W. A crosscut adit was driven a distance of 862′ through slate and at this distance intersected the vein mentioned above, at a depth of 500′ below the surface. Drifts were driven on the vein from the crosscut for a distance of 600′ north and 100′ south, and a raise was made 100′ above the tunnel level. The crosscut adit was continued for a distance of 340′ beyond the main vein and it is estimated that it will have to be extended 300′ farther before No. 2 vein is encountered. In all there are five parallel veins on the property.

The veins occur in the Calaveras slates and probably belong to the vein system worked in the Spanish mine to the north. The ore shoot developed in the Lady Oakman vein is about 350′ long and 4′ wide and the ore is said to mill $4 per ton in free gold. In the Spanish mine the ore was very low grade, averaging $1.50 per ton, but owing to ideal conditions this ore was worked at a profit. Owing to cheap power, timber and labor and to the fact that the ore could be quarried, the working costs were about 75¢ per ton.

The average mining and milling costs at the Gaston mine were in 1906 about $3 per ton. The Oakman mine is equipped with a complete 5-stamp mill, blacksmith shop and necessary buildings. Power is obtained from the South Fork of Poorman's Creek. It was last worked in 1907.

**Ocean Star Mine** (Cooley). Owner, Mrs. L. W. Cooley, Washington, Nevada County, California. Bonded to Columbia Consolidated Mines Company, and operated in conjunction with Columbia mine.

Location: Washington Mining District, Sec. 9, T. 17 N., R. 11 E., 2 miles east of Washington. Elevation 3000'.
Bibliography: U. S. Geol. Survey Folio 66, Colfax.

The Ocean Star mine is situated on the south side of the South Yuba River. There are three locations known as the Ocean Star, Hanley and South Hanley, which have an area of 60 acres and cover 3000' along the strike of the lode. The mine has been worked at intervals, but is only superficially developed. A tunnel at an elevation of 35' above the river, has been driven on the vein for 160 feet. From this a winze has been sunk on the vein to a depth of 80' and drifts have been driven 150' north and 65' south. The quartz vein varies in width from 1' to 8' with an average width of 2½ feet. Its strike is north and the dip 85° E. Water for power is obtained through one mile of 8" and 12" pipe from the South Yuba Development Company's ditch. The property is equipped with 10" x 12" Giant compressor, 8" x 10" air hoist and a 4-stamp mill, together with the necessary buildings. Six men were employed in August, 1914. Shaft to be deepened and mill capacity increased.

In the latter part of 1915, the property was taken over by Columbia Consolidated Mines Company and a winze was being sunk from the tunnel. The orebody had widened out, and the mill was running two shifts on good ore.

(Work was stopped in August, 1918, by the above company, because of the high price of labor and material, and the difficulty of getting supplies.)

**Omaha.** (See Empire West Mines.)

**Ophir.** (See Empire.)

**Orleans.** (See Sultana.)

Location: Grass Valley Mining District, Sec. 1, T. 15 N., R. 8 E., and Sec. 35, T. 16 N., R. 8 E.

**Orleans.** (See Mountaineer.)

Location: Nevada City Mining District, Sec. 13, T. 16 N., R. 8 E.

**Oro Fino.** (See Sultana.)

**Oro Grande.** (See under Copper.)

**Osborne Hill Mine.** (See Empire.) Owner, Empire Mines and Investment Company, 375 Sutter street, San Francisco.

Location: Grass Valley Mining District, Sec. 1, T. 15 N., R. 8 E., 2 miles southeast of Grass Valley. Elevation 2900'.
Bibliography: Cal. State Min. Bur. Repts. XII, page 198; XIII, page 259. U. S. Geol. Survey Folio 29, Grass Valley Special.

The Osborne Hill mine, which belongs to the Empire Company, was worked in early days. From 1852 to 1870 it was developed to an inclined depth of 400' and a large amount of rich ore is said to have been produced. Laterally the vein was productive for a distance of 800' north and 200' south of the shaft. The mine was idle from 1870 to 1894, when it was again reopened. At this time it was equipped with a 20-stamp mill and the shaft was sunk to a depth of 600', but after a few years of operation it was closed.

The property consists of the Osborne Hill, Phoenix, and Woodberry patented claims, about fifty acres in area, which control 2500' along the strike of the Osborne Hill vein system.

The Osborne Hill vein occurs in diabase-porphyrite; in some places, however, the wall rock is said to be a fine-grained siliceous clay or sandstone. The general course of the vein is N. 20° W., but there are many local undulations and numerous branches. At the surface, the dip was about 30° W., but as depth was attained the dip increased to 45°. On the 500' level the vein is said to have been from 3' to 4' in width and to have had an average value of from $30 to $40 per ton. The foot-wall is well defined and the wall rock near the vein, and that included in the vein filling, contains the alteration product sericite together with smaller amounts of calcite. As is characteristic of the southern end of the Osborne Hill vein system, the ore carries besides free gold, and pyrite, arsenopyrite and chalcopyrite with some sphalerite and galena. The sulphide content is about 2% of the ore.

**Osceola Mine.** Owners, Chas. Singleton, Rough and Ready, Nevada County, California; H. C. Shroeder, Nevada City, California. Bonded to E. W. Tarr, Grass Valley.

Location: Rough and Ready Mining District, Secs. 25 and 30, T. 16 N., R. 7 E., 4 miles west of Grass Valley. Elevation 2000'.
Bibliography: Cal. State Min. Bur. Rept. XIII, page 259. U. S. Geol. Survey Folio 18, Smartsville.

The Osceola Consolidated property is composed of the following locations: Osceola, Osceola Extension, Live Oak, Florence, Diana, Fippin, Fippin Extension, Yellow Pointer, Yellow Pointer Extension, Moss Agate, Moss Agate Extension, Aurora and Crown. The claims cover an area of 240 acres and a length of 3000' along the strike of the lode.

The vein, which is located on the south side of Squirrel Creek, strikes N. 70° W. and dips 60° N. It varies in width from 18 inches to 4 feet with a little gouge on either wall. The quartz carries heavy free gold, some chalcopyrite, and galena, and this and other veins of the district have been noted for their rich pockets of 'specimen ore.' Country rock is amphibolite and gabbro.

Most of the development work so far has been done on the Osceola claim. A tunnel has been driven on the vein a distance of 460' at a maximum depth of 150' below the surface, and for a distance of 200' from the mouth the vein was stoped to the surface. Some of the ore is said to have run $30 per ton. At a point 300' from the mouth of the tunnel a raise was put through to the surface a distance of 100 feet. The Osceola vein has been traced on the surface for 800' by shallow shafts and open cuts. Two men were working on the property in 1914. The Kenosha and California mines are in the same district.

**Oustomah Mine** (Pennsylvania and Eddy Mines). Owners, E. J. Morgan et al., Nevada City.

Location: Nevada City Mining District, Secs. 1 and 12, T. 16 N., R. 8 E., 1 mile northwest of Nevada City. Elevation 2700'.
Bibliography: U. S. Geol. Survey Folio 29, Nevada City Special.

The Oustomah property embraces the Pennsylvania patented quartz claim (100' x 2600'), the Eddy patented quartz claim (100' x 1200') and the Pennsylvania patented placer claim, a total area of 168 acres. These claims cover a distance of 4000' along the strike of the vein.

The vein was discovered during hydraulic mining operations in early days. It outcrops in the bed of one of the Old Neocene Rivers and is covered by gravel, which is in turn overlain by a volcanic capping of rhyolite and andesite. The bedrock of the gravel deposit and the wall-rock of the vein are granodiorite. In early days an inclined shaft was sunk to a depth of 500', and drifts were driven for a distance of 400' north and south. After lying idle for many years the property was reopened and the shaft sunk to a depth of 1045' on the incline. Levels and drifts have been driven as follows: No. 3 level (246' inclined depth below collar of shaft); No. 4 level (338' inclined depth) 300' north and 300' south; No. 5 level (431' inclined shaft) 1300' north and 600' south; No. 6 level (529' inclined depth) 720' north and 300' south; No. 7 level (621' incline depth) 1200' north and 720' south; No. 8 level (718' inclined depth) 600' north and 600' south; No. 9 level (819' inclined depth) 420' north and 1100' south; No. 10 level (938' inclined depth) 30' north and 420' south. On No. 4 level south of the shaft a crosscut was driven 85' into the foot-wall of the Oustomah and other veins are reported to have been encountered.

The well-defined walls of the vein are from 3' to 6' apart, and the filling is composed of from 1' to 2' of quartz and the rest of altered granodiorite. In some cases the feldspars of this brecciated granodiorite have been dissolved, leaving the residual quartz in a honey-comb structure. The quartz within the ore shoots carries free gold and from 5 to 8 per cent of sulphides, composed for the most part of pyrite with galena and zinc blende. The sulphides are said to run $100 per ton. The general strike of the vein is N. 25° W. and the dip above the 900' level is 45° NE. Below this level the vein flattened and the inclination of the shaft was changed to 20°. The vein was even flatter than this and at the bottom of the shaft it is on the hanging wall side. The orebodies are irregular in shape and occurrence. Ore has been stoped both north and south of the shaft above the 700' level, but below this level the vein is said to 'tighten.' On the No. 5 level 360' north of the shaft a 'split' in the vein was encountered, one branch running southwest and the other nearly south. A drift was driven on the south vein and some ore was stoped. On the 700' level the ore shoot was 300' in length and had a raise of 30° S. On the north branch on this level ore was 'irregular and bunchy.' No data as to tonnage or production could be obtained.

The mine is equipped with a Fairbanks-Morse 2-stage 380-cu. ft. compressor (or 14″ by 9″ by 10″) driven by a 75-horsepower electric motor; Fairbanks-Morse electric driven, 4½″ by 6″ triplex pump; single drum electric hoist (37 horsepower); blacksmith shop and 10-stamp mill (950 lb. stamps) with 4 Johnson concentrators. The mill was being operated only at intervals in 1914, and 13 men were employed at the mine. At this time the property was under bond to the Grizzly Ridge Mining Company of Los Angeles, H. D. Staley, manager. Later in the same year the property reverted to the owners who have since kept the mine partly unwatered.

Twenty men were employed early in 1915, principally doing tribute work, and the stamp mill was crushing ore from the 400', 500', 600' and 700' levels, the lower levels being ready for extraction of ore. The shaft was being repaired, new track was being laid, 1500' of cable was added to the underground equipment and four new drills were installed.

[The mine has been idle since 1916. The 10-stamp mill has been in operation, however, since the spring of 1918, crushing chromite ore.] (See Chromite.)

**Paine Claim.** Owners, H. Paine and M. Paine, North Bloomfield.
Location: North Bloomfield Mining District, Sec. ?, T. 17 N., R. 9 E., near Lake City.
Bibliography: U. S. Geol. Survey Folio 66, Colfax.

This property consists of one group of nine locations, containing about 180 acres. A shaft was sunk to a depth of 55' on a porphyry (diabase) dike. The associated quartz seams and stringers carry coarse free gold and it is reported that $1,200 in specimen ore was taken from this shaft and the 40' of drifts therefrom. A flow of water caused the abandonment of the work. Only assessment work was being done in 1914.

**Parr.** (See Pennsylvania.)

**Peabody Mine.** Owner, Golden Center of Grass Valley Mining Company, Grass Valley and Sacramento.
Location: Grass Valley Mining District, Sec. 27, T. 16 N., R. 8 E., just outside city limits of Grass Valley. Elevation 2450'.
Bibliography: Cal. State Min. Bur. Repts. XII, page 198; XIII, page 260. U. S. Geol. Survey Folio 29, Grass Valley Special.

The Peabody mine lies just north of the Gold Hill property of the North Star Mines Company. It has been worked at various times since its discovery in early days, the last time during the period from 1890 to 1893. A shaft was sunk on the vein to a depth of 400' and drifts were run both north and south. The general strike of the vein is north and the dip is 30° W. The vein occurs near the contact of the granodiorite and diabase, passing from one rock to the other. Quartz is said to have averaged 18" in width and carried coarse free gold, with 2% of sulphides. There is no equipment on the property at the present time.

The mine was purchased by the Golden Center of Grass Valley Mining Company, in June, 1915, and will be developed from the Golden Center inclined shaft.

Cross-cutting from the Golden Center shaft in the direction of the Peabody vein was suspended when a heavy flow of water was encountered. Continuation of the work may be expected as soon as the Golden Center shaft is unwatered.

**Pennsylvania Mine** (includes the W. Y. O. D.). (See Empire.) Owner, Empire Mines and Investment Company, 375 Sutter street, San Francisco.

Location: Grass Valley District, Sec. 34, T. 16 N., R. 8 E., 1 mile southeast of Grass Valley. Elevation collar of shaft 2559'.
Bibliography: Pennsylvania, Cal. State Min. Bur. Repts. X, page 383; XI, page 276; XII, page 198; W. Y. O. D., Cal. State Min. Bur. Repts. VIII, page 435; X, page 379; XI, page 268; and XIII, page 267.

The combined Pennsylvania property has, since its acquisition and systematic development by the Empire Company, become a successful

Photo No. 33. Hoist and mill buildings of the Pennsylvania Mine, now owned by Empire Mines and Investment Company.

rival in production and future possibilities to the other famous mines of the Grass Valley district.

The Pennsylvania and W. Y. O. D. interlinked veins belong to the complex Osborne Hill system of west dipping veins, with northerly strike, of which the Empire lode is the best known. The Telegraph W. Y. O. D. vein on the surface appears 2800' west of the Ophir Hill Empire vein as a narrow stringer only a few inches in width, but below the 400' level it rapidly increased in width and gold content. North of the W. Y. O. D. shaft the vein outcrops in the granodiorite, while to the south it lies in diabase porphyry. In the deeper levels most of the workings are in the granodiorite, as is also the case in the Empire and North Star mines. The strike is north and the dip is 32° W. The Pennsylvania vein lies about 500' west of the W. Y. O. D. and while it parallels the W. Y. O. D. on the surface, both veins striking in a general north direction, recent development in the lower levels shows that the Pennsylvania has a variable strike, the southern end trending southeast while the northern end bows to the northeast. The dip of this vein near the surface is 20°, but in the lower levels it steepens to 40°. It averages from 12" to 18" in width with a maximum width of 4 feet; the W. Y. O. D. varies in width from a seam to 3', probably averaging 18". The Pennsylvania vein outcrops in granodiorite and can be traced on the surface for a distance of 1000' north of the Pennsylvania shaft, but it does not outcrop south of the shaft. The granodiorite wall rock of these veins is fresh or only slightly altered.

The Pennsylvania claim, which has a length of 2900' and a width of from 200' to 400' was not worked as early as many other mines in the district, nor as extensively, and in 1890 development work consisted only of an inclined shaft 345' in depth, 950' of drifts and a 500' drain tunnel. In 1888 the mine was closed owing to the 'splitting up' of the vein, but in the summer of 1890 it was reopened and prospecting work was done until 1894, during which year a considerable amount of gold was produced. The ore carries fine and coarse free gold, and 2% of sulphides, consisting of pyrite, galena and sphalerite. The ore shoots, as is characteristic in the Grass Valley district, form irregular areas in the plane of the vein, which, as a rule, is small and low grade outside of the shoots. By 1898 the Pennsylvania shaft had reached a depth of 700' and 3000' of drifts had been run on the vein. During the year 1899 a new 20-stamp mill was installed. In February, 1900, a lawsuit was instituted against the Pensylvania company by the Grass Valley Exploration Company (W. Y. O. D.) relative to extralateral rights and in August, 1900, a counter suit was brought by the Pennsylvania against the W. Y. O. D. These suits were bitterly contested by the best legal and engineering talent until July 28, 1902, when a final decision was rendered in favor of the Pennsylvania company, which was awarded, in lieu of damages, all the property belonging to the Grass Valley Exploration Company, namely, the W. Y. O. D. Telegraph, Kate Hayes, Pan, Sims, Grant, Nuttall, and other mineral rights.

Owing to the fact that the outcrop of the W. Y. O. D. ('work your own diggings'), vein was only 3″ to 6″ in width, it was worked but superficially until 1888 when it was developed by a company of local men and good ore was encountered at a depth of 400 feet. As development proceeded below the 400′ level the vein began to widen and the value of the ore increased; at 500′ the vein averaged 10″ and the ore yielded $18 to $20 per ton in free gold; at a depth of 620′ the ledge increased in size to 2′ and at the 700′ level it was 2½′ in width and had a value of $50 per ton. The percentage of sulphides from this ore shoot, consisting of pyrite, galena and zinc-blende, varied from 2% to 3% and averaged from $250 to $300 per ton in value. The mine produced during the four years 1890, '91, '92 and '93, respectively, $26,000, $53,500, $108,700 and $143,360. By 1896 the shaft had been sunk to a depth of 1400′ and drifts had been extended a maximum distance of 600′ south and 700′ north of the shaft. In 1899 the shaft was 1600′ deep and a 20-stamp mill was in operation, 65 men being employed in the mine. Following the settlement of the suit, the Pennsylvania operated both properties but, owing to the impoverishment of the company by the costly litigation, funds were lacking for systematic development of the mine in excess of stoping requirements, and for the next ten years the hand-to-mouth method of working resulted in an average recoverable value of $8 to $8.25 per ton and a working cost of from $8.30 to $8.50 per ton, or a net loss of from 5¢ to 25¢ per ton milled.

In 1912 the mine was examined by a competent mining engineer who rightly rendered an adverse report, as the price asked was $50,000 with a large cash payment and the examination showed practically no ore in sight. Notwithstanding this poor showing, Mr. Starr again showed his faith in the mines of Grass Valley by taking over the Pennsylvania properties. He inaugurated a campaign of energetic development, which in the past had brought the Empire mine from the worked-out class to one of the best dividend-paying properties in California. The result of this business-like method of mine exploitation has been that in three years enough gold has been taken from the Pennsylvania mine to pay all the abnormal working and equipment costs, to make all the payments on the purchase price of the mine as they became due, and to yield, in addition, a handsome surplus.

In 1912 the exploration on the 1400′ level of the Pennsylvania was carried on from the W. Y. O. D. shaft as the Pennsylvania shaft at this time was only 1100' in depth. The first work undertaken by the new management was the completion of the shaft from the 11th to the 14th level by means of a raise which was put through in the record time of one month. Incidentally it may be remarked that this work had been contemplated by the former management for many years, but it was considered too long and costly a piece of work to undertake and all the ore and waste from the 1400' level was therefore handled in a roundabout way through the W. Y. O. D. shaft. The Pennsylvania shaft was continued on the vein to a depth of 1750', from which point it has been continued at an angle of 30° in the granodiorite foot-wall of the vein to a depth of 2600 feet. Since the completion of this work the W. Y. O. D. shaft has been abandoned. In the old workings of the Pennsylvania above the 1100' level, the drifts were run at intervals of 100', but below this level the drifts are from 300' to 380' apart.

Recently crosscuts have been completed from the 3400' and 4600' levels of the Empire mine to the Pennsylvania vein. The crosscut started from the 4600' level, intersected the Pennsylvania vein at a distance of 935' from the Empire shaft, which at this level is located 450' in the hanging wall of the Empire vein. This crosscut cuts the vein at a depth of 3400' on the dip and the vein is here reported to be of average width and the ore of exceptional richness. These recent developments again prove that the small fissures of the Grass Valley district are permanent and although barren horizons or zones are bound to be encountered between ore shoots, systematic development work will usually result in opening up other pay shoots. The main ore shoot now being worked was discovered by the former owners in a winze from the 1400' level, but it was not until this shoot was opened from the 1700' level by the Empire company that bonanza ore was encountered.

The Pennsylvania shaft is equipped with a 50-horsepower electric hoist, an 800-cubic foot Laidlaw-Dunn-Gordon compressor, and 2 Taylor pumps which are handling 600 gallons per minute from the 1100' level against a head of 647 feet. The metallurgical plant consists of a 20-stamp mill (1200-lb. stamps) which crushes 60 tons per day, 6 Frue vanners, and a 100-ton cyanide plant which treats the tailings only, the sulphides being treated at the Empire cyanide plant.

[In September, 1918, the Pennsylvania mill was in full operation on ore which was being hoisted through the Pennsylvania shaft from five or six levels above 2600 feet inclined depth, which is the present depth of that shaft. Below 2600 feet, the vein is being worked through crosscuts driven from the Empire shaft to the Pennsylvania vein at 3400 and 4600 feet inclined depth, as mentioned elsewhere. It was originally intended to hoist all the ore from the Pennsylvania workings through the Empire shaft, but it was found that the combined tonnage could not be properly handled through one exit, so it was decided to continue hoisting ore mined in the upper levels from the Pennsylvania shaft. A surface electric tram, now under construction, will be used when completed to haul ore from the Pennsylvania shaft to the Empire mill, and the Pennsylvania mill will then be abandoned.—*C. A. Logan.*]

**Phœnix** (Sneath & Clay).   Owner, J. A. McKennie, Los Angeles.
  Location: Nevada City Mining District, Sec. 11, T. 16 N., R. 9 E., 1 mile south-east of Nevada City.
  Bibliography: U. S. Geol. Survey, W. Lingren, Prof. Paper 73, pages 125–132. U. S. Geol. Survey 17th Ann. Rept., pt. II, pages 1–262, 1896. U. S. Geol. Survey Folios 18 and 29, Smartsville and Nevada City.

This property consists of one patented claim comprising an area of 20 acres. It was a large producer in early days when operated by Sneath and Clay.

The Sneath and Clay vein is a short vein in granodiorite. Rich quartz specimens found in the placers just below it resulted in its discovery in 1862. The mine was worked quite extensively up to 1865, producing about $200,000 and yielding ore which assayed from $32 to $180 per ton. It was also worked from 1865 to 1867, yielding a fair profit, the ore at times carrying $40 per ton. Soon afterward the mine was shut down.

In January, 1915, it was reported to have been reopened. Later in January, 1916, J. A. McKennie of Los Angeles took over the property.

The vein has a flat westerly dip of 23° and the incline is run down along the ledge. It is irregular in size, but averages something over a foot in width, with small, but rich, pay shoots. Under the new management the shaft will be deepened and additional equipment installed. Unwatering of the lower levels will begin at once.

**Sneath-Clay.**   (See Phœnix, above.)

**Pine Hill Mining Company.**   (See under Copper.)

**Pittsburg Group** (Gold Flat-Potosi).   Owners, Bert and Chas. Schlesinger, Mills Building, San Francisco.
  Location: Nevada City Mining District, Secs. 13 and 24, T. 16 N., R. 8 E., and Secs. 18 and 19, T. 16 N., R. 9 E., 1½ miles south of Nevada City. Elevation 2800′.
  Bibliography: Cal. State Min. Bur. Repts. XII, page 189; XIII, page 245. U. S. Geol. Survey Folio 29, Nevada City Special; Raymond's Report of 1872, Mining Resources West of the Rocky Mountains, pages 127–130.

The consolidated Pittsburg and Gold Flat properties comprise an irregular area of about 106 acres; including the Pittsburg, Pittsburg North Extension, Gold Flat and Potosi patented claims. The Pittsburg vein was discovered in 1851 and worked at intervals until 1862, yielding approximately $200,000. From 1862 to June, 1879, the mine was a profitable producer and is credited with a production of $900,000 during that period. The mine was again worked intermittently from 1879 until taken over by the present company in 1909, but no record of the production during these years is extant. In 1909 the present company installed new equipment and the shaft has been sunk 625', making a total depth on the vein of 1625 feet. The development work accomplished since 1909 has been confined, almost entirely to the horizon between the 1000' and 1600' levels. Above the 1000' level drifts were run every 100', the maximum distance explored being 700' north and 700' south, on the 700' level. In the new work below the 1000' level drifts were driven on the 1100' level 300' north and 300' south of the shaft; on the 1300' level 300' north and 400' south; on the 1500' level 300' north and 350' south and on the 1600' level 400' south of the shaft. On the 1100' level crosscuts were driven 100' into both the hanging wall and foot-wall of the vein, without result.

The Pittsburg vein strikes N. 45° E. and dips 43° SE., averaging 15'' in width with a maximum of three feet. The enclosing walls are hard diabase and there is comparatively little gouge on either wall. As is characteristic of the ore shoots of the Nevada City and Grass Valley districts, the pay shoots in the Pittsburg vein are irregular in shape; above the 300' level the vein was stoped from a point 400' north to 700' south of the shaft; below the 300' level this orebody appeared to divide into two shoots, one lying north and the other south of the shaft. These gradually decreased in size and the ore decreased in value as depth was attained. Apparently a barren zone was developed below the 700' level and it is the opinion of those operating the property that the ore recently developed on the 1500' and 1600' levels is the top of a new ore shoot. Such a supposition is certainly within the bounds of possibility as in a parallel case in the Empire mine after passing through a barren zone from the 1300' to the 2100' level, new orebodies were discovered which have since made this mine one of the most profitable gold mines in California. Only deeper exploratory work can decide whether such a condition exists and the past production of this property warrants further development work laterally as well as in depth.

Post-mineral faulting of the vein has occurred to a greater extent at the Pittsburg than in any other mine in the district. Numerous small normal faults with a general easterly strike and steep northerly dip intersect the vein, but the greatest displacement does not exceed 40 feet. Quartz is the principal gangue mineral. Siderite and calcite also occur in the vein in small amounts and specimens of calcite have been taken out which contained coarse gold. The ore from the upper levels of the Pittsburg mine is said to have averaged $60 per ton, but the only 'pay ore' stoped below the 700' level, was taken from a shoot about 160' in height, first developed on the 1500' and later on the 1600' level. The ore from this shoot which was 15'' in width averaged only

$8 per ton, but samples taken on the 1600' level have assayed as high as $90 per ton.

The Gold Flat or Potosi vein lies about 1000' west of the Pittsburg vein, strikes nearly north and dips 40° E. It is enclosed in hard diabase and averages about 15" in width. This vein has been developed by two inclined shafts—the Gold Flat shaft 300' deep at the northern end of the vein, and the Potosi shaft which was sunk in early days to a depth of 400 feet. The two shafts are connected by a drift on the vein, 800' in length, driven from the 200' level of the Gold Flat shaft. All of the ore above this level has been stoped. The 300' drift from the Gold Flat shaft had only been driven a distance of 150' south at the time the mine was closed, about 1896. The main ore shoot on the 200' level extended from the Potosi shaft 300' north and therefore the south drift on the 300' level had at least 300' further to go to encounter the main ore shoot. Three short ore shoots were opened on the 200' level, one 80' in length at the Gold Flat shaft, and two between this and the main Potosi shaft. The ore from the 200' level is said to have averaged better than $10 per ton.

The Gold Flat vein is crossed by the same system of post-mineral faults that were encountered in the Pittsburg mine. The throw measured along the drift is about twenty feet. The Gold Flat vein can be prospected from the Pittsburg shaft, at any desired level by driving an 1100' crosscut, which can be most economically driven by following the main east and west fault fissure.

The mining equipment on the Pittsburg property consists of a double-drum electric hoist capable of sinking to a depth of 2500'; an 850-cubic foot Giant duplex compressor driven by a 150-horsepower motor; a 10" Cornish pump which lifts the water from the 1300' level and a Ward drill sharpener. This equipment is housed in a corrugated iron building. There is a complete 20-stamp mill (1100-lb. stamps) on the property, which was being operated in 1914.

[Since the above was written, a crosscut has been driven 1500 feet from the Pittsburg shaft at the 1300-ft. level, along a cross-fissure in the direction of the Gold Flat vein. The latter vein was not encountered, although it had been estimated that it would be reached in 1100 feet. A winze has also been sunk 104 feet from the 1600-ft. level and some drifting done, but very little ore has been milled in the past four years. Idle in September, 1918.—*C. A. Logan.*]

**Polaris Claim.** Owner, Dr. G. E. Chappel, Grass Valley, California.
Location: Grass Valley Mining District, Sec. 11, T. 15 N., R. 8 E., 3 miles south of Grass Valley. Elevation 2000'.

The Polaris patented claim of 7 acres adjoins the Norambagua property on the North. Little work has been done on this prospect, and it was idle in 1914.

**Polar Star Claim.** Owner, Polar Star Gold Mining Company; Angelo Byrne, president, Department 12, Superior Court, San Francisco, California; W. W. Byrne, secretary, Commercial Club, Salt Lake City, Utah.

Location: Grass Valley Mining District, Sec. 4, T. 15 N., R. 8 E., 1 mile southwest of Grass Valley. Elevation 2400'.

This property consists of 120 acres of patented agricultural land lying one-fourth mile west of the North Star mine. The vein outcrops in a narrow belt of Calaveras slate near its contact on the west with gabbro-diorite. The general strike of the vein is north and it is said to average 18" in width. It has been only superficially developed. No work was being done in 1914.

Taken under bond by Chicago, Colorado and Utah people, in September, 1915. Cleaning out the 216' shaft will start at once and equipment will be installed capable of operating to a depth of 2000'. Oscar Coflin, superintendent. Ten men employed and number will be increased as soon as living quarters are provided. Electric hoist and pump installed. Idle in September, 1918.

**Potosi.** (See Pittsburg.)

**Powning Claim.** Owner, A. F. Brady, Coalinga, California.

Location: Grass Valley Mining District, Sec. 26, T. 16 N., R. 8 E., just outside northern limit of Grass Valley townsite. Elevation 2500'.
Bibliography: U. S. Geol. Survey Folio 29, Grass Valley Special.

This property consists of one patented claim of 20 acres. The property adjoins the Coe mine on the east and is supposed to contain the eastern extension of the Coe vein, which strikes N. 70° W. and dips 60° N. No work has been done on the Powning property for many years.

**Premier Claim.**

Location: North of Grass Valley.

Development work was being pushed on this property during the summer of 1914 under the direction of Jess R. Butler, superintendent. In May the working face was in 1800', having cut the vein at about 950'. Some rich quartz stringers were cut in a crosscut, and some rich ore was opened in a raise. Enough ore was developed to supply a 5-stamp mill for a considerable length of time. The erection of a mill is contemplated. There is said to be good gravel on the property. Idle in 1918.

**Providence.** (See Champion.)

**Prudential Mine** (Perrin and Slate Ledge). Owner, Prudential Gold Mining Company; Geo. W. Root, 2239 Atherton street, Berkeley, California.

Location: Grass Valley Mining District, Sec. 10, T. 15 N., R. 8 E., 4 miles south of Grass Valley. Elevation 2000'.
Bibliography: U. S. Geol. Survey Folio 18, Smartsville.

This mine is on agricultural patented ground situated on both sides of Wolf Creek, about four miles south of Grass Valley. It has been worked at intervals since early days. The vein which outcropped on the south bank of the creek, has been developed by an 1800' tunnel; 900' from the portal a winze was sunk 300' on the vein which has a northwesterly strike and a dip of 30° SW. This work had been done prior to 1896, but since that time considerable work has been accomplished. The mine was not operating in 1914–1915, and no information could be obtained regarding recent operations. The vein averages 18″ in width and the ore above the tunnel level is said to have run between $15 and $20 per ton. For a distance of 1600' from the mouth of the tunnel the wall-rock is diabase, but beyond this point the vein passes into the area of granodiorite in which the Norambagua and other mines of the Grass Valley district occur.

In December, 1915, the mine was secured by Geo. W. Root, and it will be operated by a group of New York capitalists. It is planned to install considerable equipment and to work along broad lines.

**Ragon Mine.**   Owner, Robert Nye, Grass Valley.

Location: Nevada City Mining District, Sec. 2, T. 16 N., R. 8 E., 2½ miles west of Nevada City by fair road. Elevation 2700'. 1½ miles north of the old Merrifield shaft of the Champion mine.
Bibliography: Cal. State Min. Bur. Rept. XIII, page 260.   U. S. Geol. Survey, W. Lindgren, Prof. Paper No. 73, pages 125–132.   U. S. Geol. Survey 17th Ann. Rept., pt.II, pages 1–262, 1896.   U. S. Geol. Survey Folios 18 and 29, Smartsville and Nevada City.

The Ragon property consists of 2 quartz locations, known as the Ragon and Ragon Extension, and placer claims. All rights to the gravel channel have been traded to the Richland Mining Company, for rights to quartz veins under their ground. There is an area of 110 acres owned, and mineral rights under 80 acres obtained from the Richland Mining Company, covering a length along the lode of 2400 feet. The surface is rolling and there is no timber on the property. The vein was first discovered in working gravel in the old Empire mine in 1856. In 1870 Ragon's incline was sunk.

The vein is developed by a 210' inclined shaft, reaching a depth of 70' on the vein below bedrock, and a drift driven from the bottom of the shaft 100' to the north. A block 50' to 100' long, 70' deep and 4' in width has been stoped. Ore is supposed to have yielded $8 per ton, recovered by amalgamation. No sulphides were saved.

The vein has well-defined walls and the quartz carries free gold, galena, pyrite and a small amount of chalcopyrite. It is a supposed northern extension of the Merrifield vein. The walls are granodiorite. The vein when worked, was 4' in width, with a northwest strike and a dip 45° NE. It has a proven length of 700', but can be traced at intervals both north and south for a distance of several miles.

The Ragon Development Company, in which R. H. Countiss of Chicago and R. Chester Turner, general manager of the Brunswick Consolidated, are interested, was organized in March, 1915, to take over and develop this property in conjunction with the Mount Auburn claims and a head frame, electric hoist and pumping plant were being installed in April, 1915. Idle in the latter part of 1918.

**Redan.** (See Sultana.)

**Red Ledge.** (See under Copper.)

**Red Ledge Mine.** Owners, R. F., T. B. and J. M. Williamson, Washington; C. M. Cole, Washington, Nevada County. Bonded to W. F. Meeks, New York.

> Location: Washington Mining District, Secs. 12 and 13, T. 17 N., R. 10 E., 1 mile south of Washington, Nevada County, 18 miles southwest to Nevada City. Elevation 2600–3600'.
> Bibliography: U. S. Geol. Survey Folio 66, Colfax.

The Red Ledge property consists of the following locations: Red Ledge, Red Ledge Extension, El Capitan, Glacier Point, Brandy Flat, Ravine, Washington, Chief of the Hills, New Year (fraction) and Triangle No. 1 and No. 2 (fractions). The total area is 150 acres, covering a length along the lode of 1 mile. The property is situated on the south side of the South Yuba River.

This mine was discovered by the present owners in 1907 and it was worked steadily from 1908 till 1914.

Development work consists of a crosscut tunnel of 50' to the ledge and a drift 400' south, reaching a depth below the outcrop of 250'. At a point 200' south of the crosscut adit a winze was sunk to a depth of 300' below the upper tunnel, and a 120' crosscut is being driven; this is in 1000' and will have to be extended 200' farther to strike the ledge.

The foot-wall of the lode is serpentine and the hanging wall Calaveras slate. Between lies an altered zone of ankerite and dolomitic material, traversed by numerous gold-bearing stringers and veinlets of quartz. The width of the lode is from 10' to 30' and mariposite. (chromium mica) accompanies the rich pockets. In the oxidized zone, below which no work has so far been done, the ankerite has been altered by weathering to an iron cap cut by quartz stringers carrying coarse leaves and plates of gold. Some beautiful specimens of leaf and crystalized gold have been taken from this mine. The general strike of the lode is N. 15° W. and the dip from 70° to 80° E. The gold occurs in 'bunches' and the so-called pay shoot is said to have been 70' in length.

Equipment consists of a 2-stamp mill, 1-drill Ingersoll-Rand compressor, 20-horsepower distillate engine, and blacksmith shop. The property was being worked in 1915, but in the spring of 1917 the owners discovered a fine deposit of high grade chromite in the serpentine near the contact, and have been busy producing that mineral from the claims during 1917 and 1918.

**Red Point Group.** Owners, J. T. Dillon et al., Nevada City.

Location: Washington Mining District, Sec. 1, T. 17 N., R. 10 E., 1 mile north of Washington, 20 miles northwest of Nevada City (N. C. N. G. R. R.) by fair road. Daily stage from Nevada City. Elevation 2750'.
Bibliography: U. S. Geol. Survey, W. Lindgren, Prof. Paper 73, pages 139–141. U. S. Geol. Survey, Folio 66, Colfax.

This property consists of three patented claims, the San Francisco, St. Patrick and Paris, a total of 55 acres, covering a length of 3000' along the lode. It is located on the point separating Poorman Creek and South Fork Yuba River, one-half mile above their junction.

The deposit is similar to the Red Ledge on the south side of the river, and is the northern extension of the same lode. The gold occurs at and near the contact of the serpentine and Calaveras slates and the vein filling is composed of ankerite and dolomite traversed by stringers of gold-bearing quartz. The strike is north and the dip 70° E. The mine was idle in 1914 and no description of the development work could be obtained.

**Republic Mine.** Owner, H. T. Dyer Estate, Alvarado, Alameda County, California. Under bond to Wm. McLean, Graniteville.

Location: Graniteville Mining District, Sec. 17, T. 18 N., R. 11 E., 2 miles southwest of Graniteville. Elevation 4500'.
Bibliography: U. S. Geol. Survey Folio 66, Colfax.

The Republic claim is 300' in width and 2300' in length along the lode. It was patented about 1876, at which time ore was being crushed in a 20-stamp mill. The mine was closed a year or two later and remained idle until reopened by the present bondee in 1905. From 1905 to 1914 the property has been worked intermittently, generally only during the summer months. Development work consists of three tunnels, which have been driven northward from the south side of the Graniteville ridge. The upper or No. 1 tunnel, 200' below the outcrop has been driven a distance of 670' on the vein and connected with the 900' National adit driven southward from the north side of the ridge.

Two hundred feet below No. 1 tunnel another tunnel was driven 250' on the vein. A raise was put through to the level above on a small ore shoot 75' in length and from 6' to 10' in width; the ore in this shoot is said by Mr. McLean to have averaged $6 per ton. Number 3 tunnel, 200' below No. 2, is now in a distance of 1200' on the vein. At 500' from the portal a raise was put through to No. 2 tunnel and another raise 1000' in connects with No. 1 tunnel 470' above. It is the intention of the present bondee to continue No. 2 adit until the south-raking National shoot is encountered. This shoot was worked by means of the National shaft from the 170' level to the surface, a distance of 320 feet. The No. 3 adit will develop this shoot at a depth of 200' below the lowest workings in the National shaft.

The Republic-National vein which can be traced on the surface for over 2 miles, occurs in the wide belt of slates lying to the west of Graniteville. Near the surface the weathered and enriched portion of the lode was stoped for a width of from 40 to 80', but on No. 3 level the vein varies in width from 2' to 12'. According to Mr. McLean the average width is 8' and the average value of the ore about $2 per ton, although some 'spots' of high-grade ore have been found. Both the foot-wall and hanging wall of the vein are slate and there is 2' of gouge on the foot-wall.

The strike of the lode is N. 3° E. and the dip near the surface is 78° E., changing with depth to 45° at No. 3 level. The ore consists of quartz, carrying free gold and from 1% to 2% of sulphides, having a value of from $50 to $90 per ton. It is claimed by Mr. McLean that the ore can be mined and milled at a cost of only $1 per ton on account of exceptionally favorable conditions, such as the low cost of water-power and timber and the opportunity offered by the topography for the developing and working of the large orebodies by means of tunnels. However, under similar conditions, at the Gaston mine, which lies a few miles south of the Republic, and which has been extensively worked, the average working costs over a period of time previous to the war have been in the neighborhood of $2.50 per ton.

The equipment consists of a 10-stamp mill, installed in 1907; Giant duplex compressor; boarding house, and blacksmith shop.

Water is used exclusively for power and is obtained from the North Bloomfield ditch under a head of 500' at a cost of 10¢ per miner's inch. The mine was being operated by lessees in December, 1914.

A few men were reported to have been working in the summer of 1918.

**Republic Group.** Owner, A. L. Gill, Grass Valley.

Location: Grass Valley Mining District, Sec. 6, T. 15 N., R. 9 E. Grass Valley is 2 miles northwest of mine. Railroad runs through property. Elevation 2800'. Bibliography: U. S. Geol. Survey, W. Lindgren, Prof. Paper 73, pages 125–132. U. S. Geol. Survey 17th Ann. Rept., pt. II, pages 1–262, 1896. U. S. Geol. Survey Folios 18 and 29, Smartsville and Nevada City.

The Republic property comprises two claims, the Republic No. 1 and No. 2, both patented and lying in the middle of a block of agricultural ground 152 acres in extent. Some work was done on these claims in early days and the shaft is said to be 200' in depth, as evidenced by a large dump, but it has caved. The extent of development work and amount of ore extracted are unknown.

The deposit consists of quartz fissure veins, which are the southeasterly extension of the Brunswick-Union Hill vein system. The country rock is diabase, covered in part by andesite.

**Roannaise.** (See Eureka.)

**Rock Roche** (Berriman Consolidated and Dromedary). (See Golden Center.)

**Rocky Glen Mine** (Glenn). Owner, C. D. Eastin Estate, Nevada City.

> Location: Graniteville Mining District, Secs. 15 and 16, T. 18 N., R. 11 E., 1 mile south of Graniteville, 31 miles northeast of Nevada City, by road. Elevation 5200'.
> Bibliography: Cal. State Min. Bur. Rept. XIII, page 245. U. S. Geol. Survey Folio 66, Colfax.

The Rocky Glen property consists of four claims, the Stacey, Illinois, Annie and Russell Ravine, the first two being patented. There are four veins, named as above.

The mine was discovered in 1852 by Milton Culbertson while gravel mining. Richard Banking and James Stacey located the Stacey claim and did the first crushing in 1866. Later, R. C. Black and Captain Irwan bought the Illinois claim from John Young and consolidated the property. It is said that the old Stacey and Illinois workings have yielded over $300,000.

Work on the Banbury and Stacey vein was started by taking out the croppings through an open cut, then by a short tunnel, and later by another tunnel 90' lower. The present drain tunnel cuts the vein at a depth of 230 feet. It was driven as a crosscut 456' to the Illinois vein and then continued 700' to where it cut the Stacey. At this point an incline on the vein was sunk to a depth of 275 feet. In the incline shaft, only the highest grade rock was taken out and no prospect work was done. The 456' crosscut tunnel cut the Illinois vein 100' below its outcrop. Both veins were stoped to the surface. In 1875 John Hippert reopened the tunnel and drifted north 5300' on the Illinois. The Annie vein has been worked to a limited extent. It is situated 100' below the mouth of the main drain tunnel, its deepest workings being not more than 95'. A 50' crosscut tunnel cuts the vein just below the surface, from which point a 300' drift was run north on the vein. A 1700' tunnel with a raise of 100' cuts the Russell Ravine vein 100' below the surface, and a shoot 90' long was stoped to the surface. A 2000' drain tunnel would crosscut and drain all of the veins.

The Stacey vein ranges in width from a few inches to 14', the mean being 3½ feet. The average yield of the ore has been $6 to $8 per ton, but bunches of 'high-grade' rock are often encountered. The vein occurs in granite near its contact with the Calaveras slates. The foot and hanging walls are well-defined with a gouge carrying free gold.

There are two ore shoots in the mine, known as the North and South pay shoots, between which intervenes a 100′ block of low-grade quartz of about $1.50 a ton in value. The quartz in this mine, often termed ribbon rock, crushes easily and is of a light blue color, with bright gold which readily amalgamates. The rock carries considerable sulphurets with some galena and silver. The vein strikes north and dips 45° E. About 400′ to the west is the Illinois vein, almost vertical but with a slight dip to the east. This vein resembles the Stacey, except in dip. A small pay shoot encountered, 50′ long by 4′ wide, averaged $12 per ton.

The Annie vein lies west of the Illinois about 450′ and near the contact of the slate and granite belt. The walls of the vein are well defined, smooth, and carrying about 2″ of clay gouge. This vein pitches east toward the Illinois at an angle of about 30°. The orebodies lie in oblong masses, the character of the rock differing but little from the other veins, except in the north and where a belt of soft 'porphyry' with thin rusty streaks of quartz was encountered. The average value of the ore in the mine, including 100 tons of porphyry, has been about $10 per ton net.

The Russell Ravine claim lies to the south of the Illinois. The vein is 200′ west of the Illinois vein, its course being at an angle to the Stacey and Illinois veins, but the intersections are in ground not yet opened. The strike is a little north of east, and it has a slight dip to the south. In general this vein resembles the others. The average width is 2½′; one shoot 90′ long has been stoped to the surface. The ore is said to have averaged about $7 per ton.

In many ways the above claims are very favorably situated for profitable working. Mining timber and fire wood are easily obtainable, since 2 sawmills are near.

Equipment consists of a 10-stamp mill (water power), blacksmith shop, change house and boarding house, but all are in a dilapidated condition.

Water power is available from the North Bloomfield ditch and was used to run the mill under a 265′ head.

**Rose Hill Claim.** Owner, Empire Mines and Investment Company, 375 Sutter street, San Francisco.

Location: Grass Valley Mining District, Sec. 34, T. 16 N., R. 8 E., situated in Grass Valley townsite. Elevation 2600′.
Bibliography: Cal. State Min. Bur. Rept. XIII, page 262. U. S. Geol. Survey, W. Lindgren, Prof. Paper 73, pages 125–132. U. S. Geol. Survey 17th Ann. Rept., pt. II, page 236, 1896. U. S. Geol. Survey Folios 18 and 29, Smartsville and Nevada City.

The Rose Hill claim of 20 acres was patented earlier than Grass Valley townsite, and is said to have produced about $100,000 in those days, but the exact amount is unknown. It was idle for some time, but previous to 1913 the tunnel was cleaned out and a few years ago $7,000 in specimen ore was extracted from a small pocket. The vein is from 4″ to 15″

wide. A 110' raise will be driven from a station near the vein to the surface; this will be used as a shaft which will be sunk on the vein below the tunnel level, and the old tunnel will be used for drainage only. The vein strikes north and dips 30° W. It occurs in granodiorite and is the southern continuation of the vein now being successfully worked in the adjoining Golden Center property on the north.

**Russell Ravine.** (See Rocky Glen.)

**Scotia Claim.** Owner, Scotia Mining Company, 100 Williams street, New York; Frederick W. Yates, Nassau street, New York (attorney).

Location: Grass Valley Mining District, Sec. 27, T. 16 N., R. 8 E., ¼ mile west of Grass Valley. Elevation 2600'.
Bibliography: Cal. State Min. Bur. Register of Nevada County Mines. U. S. Geol. Survey Folio 29, Grass Valley Special.

This property consists of one irregular-shaped claim, of about 10 acres. In 1881 a shaft was sunk on a small vein to a depth of 300 feet. The strike is northwest and the dip is about 40° SW. No ore was found and the mine has remained idle for many years.

**Sebastopol.** (See Sultana.)

**Shamrock Claim.** Owner, P. Riley, Grass Valley, California.

Location: Grass Valley Mining District, Sec. 14, T. 15 N., R. 8 E., 4 miles south of Grass Valley. Elevation 3000'.
Bibliography: U. S. Geol. Survey Folio 18, Smartsville.

The Shamrock claim is about 10 acres in area. The vein, which strikes northeast, is small, averaging about 15 inches. Only superficial development work has been done and the mine was idle in 1914.

**Sharpe Mining Property.** · Owner, R. Sharpe, Nevada City.

Location: Nevada City Mining District, Secs. 17, 18 and 20, T. 16 N., R. 9 E., 2 miles east of Nevada City, by good roads. Elevation 2700-3200'.
Bibliography: U. S. Geol. Survey, W. Lindgren, Prof. Paper 73, pages 125-132. U. S. Geol. Survey 17th Ann. Rept., pt. II, pages 194-196, 1896. U. S. Geol. Survey Folios 18 and 29, Smartsville and Nevada City Special.

The Sharpe Mining property consists of quartz and placer locations, containing a total area of about 120 acres. The Greenman location extends along the Greenman vein for 1500', the Kellogg location is 1400' in length with 300'-600' of the Kellogg vein defined by croppings, the Red Rose claim extends 1400' along the westerly extension of the St. Louis vein. The Ben Harrison claim extends across the south end of the Greenman claim and contains 1100' of the Butterfly and Ben Harrison ledges. The Sharpe claim and the Glencoe Extension cover 500' and 1500', respectively, along the Glencoe-Orleans vein. The Enterprise claim extends 1600' along the Enterprise ledge, and the Hickson claim contains two outcrops of the Hickson ledge of from 250'-500' each.

This property was located and worked superficially in early days, but the veins on the property have never been prospected to any great extent, and at present the property is idle.

Within the Red Rose claim, a vein coursing on the projected strike of the St. Louis vein, but of reverse or south dip, has been explored by surface cuts and shafts on the croppings. A tunnel about 80' below the surface has prospected the Greenman vein for 400' to 600', and 150' to 200' north from the mouth of the tunnel an incline sunk about 200' on the vein developed a small shoot which yielded from $35 to $60 per ton. Three inclines have been sunk on the Enterprise ledge at intervals of 200' to 500'. From the northerly and more recent workings ore containing as high as $60 per ton has been extracted. A tunnel was run for 400' above the Beckman vein, then sunk 15' below to the vein. A lower tunnel, under the ledge, was run for 750' and then raised to the ledge. On the Glencoe, after a few tons of rock were taken from a shaft 75' deep, a drift was run west to the end lines of the Glencoe Extension. The Hickson vein was followed for 800' to its intersection with the Enterprise, then the Enterprise was followed for 50' to the shaft. Ore from the Hickson varied between $7 and $25.

All of the veins on this property are quartz-filled fissure veins containing free gold, pyrite and galena, arsenopyrite and zinc-blende. The Glencoe-Orleans vein is a fissure zone in Calaveras slate and while this vein formation is often several feet in width, the quartz in the vein does not exceed 2'-4' in thickness. Where prospected ore from this vein has yielded from $6 to $12 per ton. The lode strikes a little north of west and dips 85° S. The St. Louis is a remarkably strong vein, striking northeast with steep northerly dip. From its easterly end the vein shows great strength, cutting all cross veins, and extends to the Sharpe and McCormick ground where it breaks into several strong stringers which continue a few hundred feet westerly. The ore, where developed, has averaged $3 to $5 per ton. The Butterfly vein varies from 12'-18' in width, strikes east, and has a nearly vertical dip. At the southeasterly end of the Canada Hill and Sharpe Mining ground, the St. Louis and Glencoe ledges are 3000' apart at their croppings and dip in opposite directions. Westerly 3600', at the west end of the Sharpe ground, the ledges are 1200' apart. Within this block across the St. Louis and Glencoe veins, is a system of strong north coursing veins, of both east and west dip. The Greenman ledge apexes at the surface 450' west of the Canada Hill croppings. This ledge can be traced by surface croppings and underground development from the Butterfly and Harrison ledges, northerly 1600' to 1700' almost up to the St. Louis cross vein. The Greenman is a fair-sized ledge dipping 30° E. and striking north. Ore taken from the one shoot developed yielded from $35 to $60 per ton. South of the Glencoe vein, the Enterprise ledge, a strong 45° east dipping ledge, continues on the projected course of the Greenman. Though larger and stronger, it may be a continuation of the Greenman fissure and vein. Ore as high as $60 per ton has been mined and milled. The Kellogg vein, 500' west and parallel with the Greenman, has produced ore which yielded $16 per ton in free gold; this vein dips 45° E.

This district is favorably situated for economical mining. Being of easy access by rail and wagon roads, freights are reasonable, water can be obtained under 300' to 400' head and electric power can be purchased at reasonable cost. By combining the property owned by the Canada Hill mines and the Nolan and Nevada Quartz mine with the Sharpe mining property, there would be a total of 175 acres covering 13,000' of known quartz veins. The result of the work so far done in this part of the Nevada City district, warrants further exploratory work.

**Sharpe Quartz and Gravel Mining Company** (Black Bear). Owner, J. J. Jackson, Nevada City.

> Location: Nevada City Mining District, Sec. 32, T. 17 N., R. 9 E., 3 miles northeast of Nevada City. Elevation 3000'.
> Bibliography: U. S. Geol. Survey, W. Lindgren, Prof. Paper 73, pages 125–132. U. S. Geol. Survey Folio 29, Nevada City Special.

This property has been relocated many times. It was last located by Jackson in 1911, as a quartz and placer mine, and since then only assessment work has been done. The gravel was worked by a tunnel thirty years ago and at that time a vein 2' wide was discovered from which quartz was taken ranging in value from $20 to $50 per ton.

**Slate Ledge.** (See Prudential.)

**Snowflake Claim.** Bonded to J. R. Butler.

> Location: Grass Valley Mining District.
> Bibliography: U. S. Geol. Survey, W. Lindgren, Prof. Paper 73, pages 125–132. U. S. Geol. Survey 17th Ann. Rept., pt. II, pages 1–262, 1896. U. S. Geol. Survey Folios 18 and 29, Smartsville and Nevada City.

The Snowflake claim was bonded to Butler in the summer of 1915. A 5-stamp mill is to be placed in commission shortly.

**South Banner Group.** Owner, John T. Morgan, Nevada City.

> Location: Nevada City Mining District, Secs. 16 and 21, T. 16 N., R. 9 E., 3 miles east of Nevada City, by fair roads.
> Bibliography: U. S. Geol. Survey, W. Lindgren, Prof. Paper 73, pages 125–132. U. S. Geol. Survey 17th Ann. Rept., pt. II, pages 198–199, 1896. U. S. Geol. Survey Folios 18 and 29, Smartsville and Nevada City Special.

The South Banner property consists of a 20-acre quartz claim and 25 acres held by mineral patent. The vein has been developed by a 400' crosscut, cutting the vein 50' below its outcrop. Some drifting was done on the vein without results.

**South End.** (See Goodall.)

**South Idaho Mine.** Owners, W. J. Rogers, Los Angeles; James McKenna, Jos. O'Keefe, G. V. and Elesia Crase, all of Grass Valley.

Under option to Gold Point Consolidated Mines, Inc.,* 1007 Crocker Building, San Francisco.

Location: Grass Valley Mining District, Sec. 26, T. 16 N., R. 8 E., 1 mile east of Grass Valley. Elevation 2600.'
Bibliography: Cal. State Min. Bur. Rept. XIII, page 264. U. S. Geol. Survey Folio 29. U. S. Geol. Survey 17th Ann. Rept. pt. II.

The S. B. C. and O. K. patented claims comprise an area of 20 acres known as the South Idaho group. The South Idaho vein lies 1500' south of and parallel to the famous Eureka-Idaho vein, thus occupying the same relative position to each other as do the Empire and Pennsylvania veins. It occurs in the area of diabase-gabbro that forms the hanging wall of the Eureka-Idaho lode, but the wall-rock of this vein is altered to a greater degree than in most of the other veins of the district. The diabase and gabbro for a width of 8' to 15' are bleached and impregnated with pyrite and calcite by hydro-thermal agencies. The vein strikes N. 85° W. and dips 70° S.

The lode has been developed to a depth of only 155 feet by the inclined shaft sunk on the foot-wall of the vein. At a depth of 60' a crosscut was driven 12' south from the shaft without encountering the hanging wall of the vein. It is reported that over $25,000 in specimen ore was taken from a rich vein of quartz from 6" to 1' in width which followed the foot-wall from the surface to a depth of from 60 to 100 feet. At that depth the stringer extended out toward the hanging-wall, as is characteristic of the veins in this district, and 'split up' into veinlets rich in coarse gold. Besides the free gold, the ore carries about 4% of sulphides, consisting of pyrite, galena and zinc-blende. The well-defined foot-wall continued on down below the split. The only drifting done on the vein was on the 100' level where a drift was driven a distance of 25' south of the shaft. A tunnel was driven a distance of 800' on the vein east of the shaft and opened a well-defined vein of quartz from 2' to 4' in width.

The property has been owned by local people for many years and they have been unable to finance the equipment and development of the mine. It was under option (1918) to the Gold Point Consolidated Mines Company, Incorporated.

**South Star Group.**   Owner, M. A. Thorson.

Location: Rough and Ready Mining District, Secs. 30 and 31, T. 16 N., R. 8 E., 3 miles west of Grass Valley. Elevation about 2500'.

The property contains four locations, the South Star and Mill Site being patented, and Sumner and Alta or McDonald, unpatented. The area is about 60 acres. There are said to be two main veins, named South Star and Sumner, and three spur veins. Strike varies from northeast to north and dip is about 60° W. The locations cover about 3000 feet along the lode, and the veins have been proven for 2500 feet on the surface.

*Since the above was written this property has passed into the possession of the Idaho-Maryland Mines Company, Hobart Bldg., San Francisco.

Several shallow shafts have been sunk on the South Star, according to the claim map of the mine, but the property is idle now. The deepest is reported to have followed the vein for 108 feet. From this shaft a level is said to have been driven north about 160 feet, from which a stope 60 feet long is supposed to have been put up 50 feet on the vein. The work was done years ago and is said to have paid well, 300 tons having been milled in a nearby custom plant after the high-grade had been picked out. The claims lie just north of the California King group, the Alta claim having its southeast corner well within the Pittsburg North Extension claim.

**South Yuba Mine** (Red Ledge). Owner, South Yuba Mining and Smelting Company; A. A. Elliger, president, Grass Valley; J. P. Maleville, secretary, Grass Valley.

Location: French Corral Mining District, Sec. 36, T. 17 N., R. 8 E., 1 mile to French Corral, thence 15 miles, by road, southeast to Nevada City. Elevation 2700'.
Bibliography: U. S. Geol. Survey, W. Lindgren, Prof. Paper 73, pages 121–125. U. S. Geol. Survey Folio 18, Smartsville.

The South Yuba property consists of 6 full claims, each 600' by 1500' and all patented. They are: The Virginia and Gold Hill, the Martha and the Edison No. 1, 2, 3 and 4. The property rises very abruptly from the South Yuba River, giving good tunnel sites. By running a tunnel near the river, 1000' of backs can be developed. The Virginia and Gold Hill claim was worked for many years by a former owner named Roberts, for the gold in the gossan capping.

Development consists of 7 tunnels. No. 2 tunnel was in 170' on Edison No. 4, and was being driven to tap the large outcrop of gossan on that claim. It would have to be driven 50' further to tap the ore shoot at 150' depth. Tunnels No. 2, 3 and 4 are all on the same level from 200'–400' apart on the Virginia and Gold Hill claim. Tunnel No. 5 was driven about 50' directly below No. 4. Tunnel No. 6 is 135' below No. 2 and is in 275 feet. The ore was struck and 10' of it is exposed. Tunnel No. 7 is a 175' crosscut located 400' below No. 6 tunnel. It will have to be driven 500' to tap the vein and give 100' of backs. In all there is about 1700' of underground work. In tunnel No. 2 a crosscut has located 20' of ore. In No. 3, 54 ft. of ore was crosscut and in No. 5 the ore is crosscut for 35'.

The ore deposit occurs in granodiorite and is impregnated with pyrite, carrying small amounts of gold and copper. In the oxidized surface zone the gossan has been enriched in gold by concentration. The deposit can be traced the entire length of the Company's property. Its course is northeasterly and southwesterly and the dip is to the west. There are three large gossan outcrops: No. 1 is on the Virginia and Gold Hill claims, No. 2 a continuation of Outcrop No. 1, is on Edison No. 1; the 3rd outcrop, on Edison No. 4 is 400' wide and 1000' long.

The deposit is richer in copper toward the hanging wall and harder and richer in sulphur toward the foot-wall, according to the owners. Assays of the low-grade ore gave 2.12% copper and $3.34 gold; the high-grade ore shows $1.92 gold and silver and 38.2% copper. Assays made of the gossan show values of from $5 to $10 per ton in gold.

The property is equipped with a 10-stamp (water power) mill and cyanide plant, connected with the mine by a gravity tramway.

Electricity can be purchased from the Pacific Gas and Electric Company and free water can be had for part of the season half a mile from the mine. Water can also be purchased at a very reasonable rate. Lumber costs less than $20 per 1000' delivered at the mine.

**Spanish Claim.** (See Champion.) Owner, North Star Mines Company.

Location: Nevada City Mining District, Sec. 2, T. 16 N., R. 8 E.

**Spanish Mine** (Spanish Ridge). Owner, F. W. Bradley, 1022 Crocker Building, San Francisco.

Location: Secs. 19 and 30, T. 18 N., R. 11 E, 3 miles southwest of Graniteville, which is 27 miles from Nevada City. Elevation 4000'.

The property consists of nine patented claims, named American, Spanish, Santa Anita, Santa Anita Extension, Pine Tree, Singleton, Grizzly, Savage and Mexican, containing in all 175 acres.

The claims cover two miles along an altered zone which contains a low-grade quartz vein four feet wide, striking north and dipping 80° W. in Calaveras slate. The gold content recovered was mostly free and ran from 50¢ to $1.25 a ton.

Two tunnels were driven, one 1500 feet long, 650 feet below the outcrop and the second 300 feet higher and 1200 feet long. Raises were put up from the ends of the tunnels and the ore in the surface zone was worked by open cuts around the raises so that gravity was fully utilized. With Chinese labor at $1.50 a day, cheap water power and plenty of cheap timber it was possible to work at a cost of from 50¢ to 80¢ a ton, it is said. The property was equipped with ten stamps and four Huntington mills, operated by water under 600' head. Capacity of the plant was about 4000 tons a month.

The property was discovered in 1883 and was worked until 1898, but has been idle lately. In September, 1918, it was said that the ten stamps were to be moved to a property near Alleghany.

**Spence Mineral Company.** (See under Copper.)

**Starr Group** (Cooley). Owner, Starr Mining Company, Grass Valley; Geo. Starr, agent, Grass Valley.

Location: Washington Mining District, Sec. 9, T. 17 N., R. 11 E., 2½ miles east of Washington, thence 20 miles, by road, to Nevada City (N. C. N. G. R. R.). Elevation 3000'.
Bibliography: U. S. Geol. Survey, W. Lindgren, Prof. Paper 73, pages 139–141. U. S. Geol. Survey Folio 66, Colfax.

The Starr property consists of 2 patented claims: The North Starr on the south side, and the Golden Gate, on the north side of the South Yuba River. The claims are 40 acres in area and cover a length along the lode of 3000 feet north and south. There is good timber on the property. The vein belongs to the Washington-Maybert-Erie fissure system, and cuts obliquely across the strike of the Calaveras slates in which it occurs. The strike is N. 10° W., the dip 80° E., and it is probably a continuation of the veins recently developed in the Columbia property to the north. No work was being done in 1914.

**Steephollow Group.** Owner, Wm. Fowler, Dutch Flat.

> Location: You Bet Mining District, Secs. 14 and 23, T. 16 N., R. 10 E., 6 miles from Dutch Flat (S. P. R. R.).
> Bibliography: U. S. Geol. Survey, W. Lindgren, Prof. Paper 73, page 144. U S. Geol. Survey Folio 66, Colfax.

The Steephollow property consists of 2 unpatented claims, one on each side of the creek. The mine is idle, only assessment work having been done for the past four years.

The vein is quartz from 4'–5' in width, containing free gold and pyrite. There are three tunnels on the vein, the longest being 160', the others 40 and 60 feet. The foot-wall is schist, the hanging wall black slate, and the vein has a proven length on the surface of 2 miles, cutting across Little Greenhorn Creek with a northerly strike and a dip of 45° W.

**St. Louis.** (See Alpine.)

**Sultana Mine** (Consolidation of Orleans, Prescott Hill, Electric, Houston Hill and Centennial). Owner, Sultana Gold Mining Company, San Francisco; J. E. Green, president, 409 Crocker Building, San Francisco; G. A. Nichols, secretary, 409 Crocker Building, San Francisco; Albert Crase, superintendent, Grass Valley.

> Location: Grass Valley Mining District, Sec. 1, T. 15 N., R. 8 E., and Sec. 35, T. 15 N., 1½ miles southeast of Grass Valley. Elevation 2600–3000'.
> Bibliography: Cal. State Min. Bur. Repts. XI, page 280; XII, page 186; XIII, pages 238, 242, 258. U. S. Geol. Survey, W. Lindgren, Prof. Paper 73, pages 125–132. U. S. Geol. Survey 17th Ann. Rept. pt. II, pages 254–256, 1896. U. S. Geol. Survey Folios 18 and 29, Smartsville and Nevada City Special.

The Consolidated Sultana property adjoins the Empire mine on the south and covers a number of veins belonging to the Empire-Osborne Hill complex, linked-vein system. The 27 claims and fractions now comprising the property are the Betsy, Biggs & Sims, Brockington, Electric, Fillmore, Houston Hill, Josephine, Kings Hill, Lizard, Madison Hill, Mills No. 1 and 2, No. 4 location, Old Houston Hill, Oro Fino, Pinnacle, Prescott Hill, Sanders, Sebastopol, South Centennial, Summit, Teddy, and Wasp, a total of 250 acres. It is situated on the top of Osborne Hill and the surrounding rolling plateau.

Portions of this property were worked in early days. The old Orleans shaft was sunk to a depth of 100' and the Houston Hill vein was worked from 1861–1870 to a depth of 300′ and produced from June, 1864, to April, 1867, a total of $500,000. The Sebastopol workings, on the north end of the Sultana vein produced $200,000, between 1856 and 1858, from a depth of 180′ on the incline. Between 1876 and 1883 the Centennial mine is reported to have produced a total of $600,000, and since 1903 the present company has taken out about $750,000.

The Electric or Sultana shaft, on the Sebastopol Electric lode, prior to 1896, had reached a depth of 400′ on the vein and later this shaft was continued to a depth of 600 feet. From a point 1500′ south of the shaft on the 600′ level, a winze was sunk on the vein for a distance of 1100 feet. Drifts were driven on the vein on the 6th, 8th and 12th levels, 800′ north and 1000′ south, and on the 16th level 250′ north. Crosscuts were also driven into the hanging wall 100′ on the 8th level, 300′ on the 12th level and 86′ on the 16th level. The shaft on the Prescott Hill ledge, worked to a depth of 300′ on the incline prior to 1867, was reopened by the Sultana company in March, 1906, and in September, 1909, had reached a depth of 1750′ on an average dip of 25° W. Drifts were run on the vein as follows: 4th level (2000′ south); 7th (3500′ south); 12th (500′ north and 1500′ south). On the 17th level a crosscut was also driven 800′ eastward into the foot-wall, and on the 7th level 1200′ in the same direction. The Orleans shaft has reached a depth of 800′ on the vein which has an average dip of 40° W. Drifts were driven on the various levels both north and south and on the 700′ level crosscuts were extended 250′ into the hanging wall and 300′ into the foot-wall at which point the Houston Hill vein was encountered. This vein was first worked from 1861–1870 by a 300′ incline shaft from which a number of levels were driven north and south. During March, 1916, the unwatering of the 800′ shaft was proceeding and sinking will continue to a depth of 1000′. Since 1903 the Sultana has driven over 12,000′ of drifts and 3000′ of crosscuts and the ore thus developed has yielded about $750,000.

All of the veins in the Sultana property occur in the diabase-porphyry with the single exception of the lower workings (1750′) of the Prescott Hill shaft, which is said to have entered the main area of granodiorite. The change in wall-rock does not affect either the size or mineral content of the orebodies. The Rich Hill vein has a strike N. 10° W. and dips 30° W. and passes through the Fillmore claim of the Sultana group for a distance of 1000′. The Ophir vein, parallel and 500′ to the east, after entering the Madison Hill claim, gradually turns to the southeast, passing through the Prescott Hill and Betsey claims into the property of the Empire Mines Company. When worked by the Orleans shaft, the dip varied from 12° to 52°, averaging 40°

W., with a width of 2′ between walls and about 8″ of quartz. The ore carries 3% of sulphurets, averaging 35 ounces of gold and 0.7 ounce of silver. The Prescott Hill ledge has an average dip, for the 1750′ developed, of 25° W.; the strike varies from north to northwest. The vein as disclosed by development work, was small, averaging only 12″, and all of the ore milled was taken from the stopes above the 700′ level. The Sebastopol Electric lode lies 700′ east of the south end and parallel to the Orleans and Prescott Hill ledge. The vein has an average dip of 35° W. An orebody 300′ in length and averaging from 2′–3′ in width, was developed above the 800′ level. The ore from this shoot was rich and the mineral content consisted of free gold, pyrite, arseno-pyrite, galena and a small amount of chalcopyrite. The Houston vein, which lies north of the Sebastopol and several hundred feet east of the Ophir-Prescott Hill vein, strikes N. 45° W. and dips 45° SW. and can be traced for 1500′ on the surface. Ore produced from the old workings of this small 8″ vein was very rich, yielding as high as $160 per ton. Later development at depth failed to open any ore shoots of commercial value.

The mechanical equipment, which has recently been moved from the Prescott Hill shaft to the Orleans shaft, consists of a Giant Duplex, 14″ x 16″ x 24″, compressor driven by a 150-horsepower motor, a double drum electric hoist (2000′) driven by a 50-horsepower motor, a 20-stamp (1250-lb. stamps), Risdon mill built in 1903, with 3 Frue and 1 Johnson vanners, and canvas plant, together with all necessary buildings, shops, etc.

Electricity is used throughout for power and is obtained from the Pacific Gas and Electric Company. Labor costs from $3.25 a day up.

The Empire and Osborne Hill mines adjoin.

[Since 1916 the Orleans shaft has been unwatered. Its total depth to the sump is 965 feet. From the 800-ft. level, a winze has been sunk 1250 feet to the Rich Hill vein, which is being explored, and shows a width of 4 feet. A raise has been put up 520 feet, within 70 feet of the surface. The Empire-Orleans vein, which narrowed to a width of 3 inches, is not being worked. So far the mill has not crushed any ore from the Orleans shaft. A 22-horsepower hoist is installed at the winze. Thirty-three men were employed in September, 1918, and the mill was idle. The mine workings develop very little water.— *C. A. Logan.*]

**Sweet Ranch.** Owner, John Sweet, Wolf post office.

Location: Spenceville Mining District, Sec. 13, T. 14 N., R. 7 E., 14 miles, by good wagon road, northwest of Auburn (S. P. R. R.). Elevation 1800′. Bibliography: Cal. State Min. Bur. Bull. 50, page 201. Folio 16, Smartsville.

This property has a total area of 210 acres.

The vein strikes northerly and dips to the east. Development work consists of two or three 10-foot holes. The gossan is iron-stained rhyolite and porphyry with limonite and hematite. Values reported from croppings are from $1.50 to $12 per ton in gold and silver.

**Syndicate Consolidated Group** (Golden Chain, Great Eastern and New Idea). Owner, Grass Valley Consolidated Mines Company.

> Location: Grass Valley Mining District, Sec. 2, T. 15 N., R. 8 E., 3 miles south of Grass Valley. Elevation 2200'.
> Bibliography: Cal. State Min. Bur. Rept. XIII, page 265. U. S. Geol. Survey Folio 29, Grass Valley Special.

The Syndicate Consolidated, which adjoins the Allison Ranch mine on the south and east, embraces the Golden Chain, Great Eastern and New Idea claims, containing 54 acres of patented land. On the Golden Chain and Great Eastern claims all of the so-called conjugate veins which strike northwest and dip 45° E. have been traced on the surface for a distance of 1500' by shallow shafts and open cuts. The southern extension of the Allison Ranch-Cariboo vein is supposed to traverse the New Idea claim, but as the vein is small, the outcrop can not be traced for any distance. This vein has yielded some rich specimen ore in the Allison Ranch mine.

**Texas Mine** (New York). Owner, Fred Ayer; W. F. Duboise, agent, Nevada City.

> Location: Nevada City Mining District, Sec. 8, T. 16 N., R. 9 E., 3 miles east of Nevada City. Elevation 2600'.
> Bibliography: U. S. Geol. Survey Folio 29, Nevada City Special.

There are four well-defined veins on this property known as the Creek, Wheel, New York and Delaware. The general strike is east and the dip varies from 20° to 50° N. These veins cross Mosquito Creek, 2000' above its junction with Deer Creek; they can be traced for a distance of from 1000' to 1500' on the surface and have been developed by tunnels driven on the vein from Mosquito and Deer creeks. The New York vein, upon which a 250' inclined shaft was sunk, is said to have a maximum width of 4 feet. All of the veins occur in the granodiorite which forms both walls. Ore was stoped having a value of from $8 to $40 and containing 5% sulphides, but no work has been done for a number of years.

The Texas, together with the Guild and Niagara mines, were taken under bond by Wm. H. Tuttle of Reno, representing Utah and Nevada people, in April, 1916. A crosscut will be run from the 600' level of the Texas to develop the Guild and Niagara.

**Tola Group.** (See under Copper.)

**Turner Group.** (See under Copper.)

**Union Mine.** Bonded to Nevada County Mines Company; D. W. Shanks, Los Angeles; J. Nelson Nevius, Los Angeles.

Location: Nevada City Mining District, Sec. 16, T. 16 N., R. 9 E., 3 miles east of Nevada City.
Bibliography: U. S. Geol. Survey 17th Ann Rept., pt. II, page 197, 1896. U. S. Geol. Survey Folios 18 and 29, Smartsville and Nevada City.

The Union mine was taken under bond by the Nevada County Mines Company, organized by Los Angeles people, in September, 1915. A gasoline engine has been installed to operate the pump and unwatering of the 225' shaft has begun.

The Union vein is located about a mile east of Canada Hill, on the north side of Little Deer Creek. It is the most westerly of the Banner Hill complex. Development consists of an incline shaft and a tunnel from the creek. At a depth of 200' on the incline the vein is said to cross without any change in its general character, from diorite into argillite. The dip is 34° E. and the width is reported to be from 1 to 4 feet. The mine was worked rather extensively from 1863 to 1867, yielding some rich ore, but mostly low-grade. Very little work has been done since.

**Union Hill Mine.** Owner, Gold Point Consolidated Mines, Inc.,* Crocker Building, San Francisco; F. W. McNear, president.

Location: Grass Valley Mining District, Secs. 25 and 26, T. 16 N., R. 8 E., 2 miles east of Grass Valley. Elevation 2700'.
Bigliography; Cal. State Min. Bur. Register of Nevada County Mines. U. S. Geol. Survey Folio 29, Grass Valley Special.

The Union Hill vein was one of the first to be discovered and worked in the Grass Valley district. In 1854 ore from the Union Hill, Cambridge-Lucky, and Greek veins was being worked in arrastras. In

Photo No. 36. Shaft house and mill of Union Hill Mine.

*Since the above was written this property has passed into the possession of the Idaho-Maryland Mines Company, Hobart Bldg., San Francisco.

1865 a mill and hoist were erected and the mine was worked at a profit until 1870, when it was closed, after having attained a depth of 300 feet. Ore from the Cambridge-Lucky vein during this period is said to have averaged $20 per ton. After remaining idle for a period of over 30 years, the mine was reopened and the shaft was sunk to a depth of 600 feet. In order to straighten the shaft it was put down in the foot-wall of the Union Hill vein and at the 600' level it was necessary to crosscut 150' south before the vein was encountered. A drift was then driven a few hundred feet east on the vein and a small body of ore was stoped between the 600' and 300' levels. The main crosscut was then continued beyond the Union Hill ledge and 500' south of the shaft the Tungsten vein was intersected. A drift was driven 1000' west on this vein and from its end a crosscut was driven into the foot-wall northwest, and at 300' encountered the Greek vein, on which a drift was extended 300' further west. The main crosscut was driven 800' beyond the Tungsten vein and there struck the Cambridge-Lucky vein. A drift was driven 1500' west on this ledge, which is said to have been 3' in width, but the ore developed had an average value of only $5 per ton. The 200' level has been driven 2400' west and 600' east and the 300' level 1500' west and 500' east. From the 200' level at a point about 900' west of the shaft a crosscut was driven 700' south where it inter-sected the Greek ledge and some good ore was stoped therefrom. On the 300' level, 1000' west of the shaft, a crosscut was driven 1300' south, opening both the Greek and Lucky veins. The latter vein it was found had been worked to a depth greater than where the crosscut intersected it. After further mismanagement the mine was closed and reverted to the owners. The property was reopened in the fall of 1914 by a San Francisco company; the shaft was continued to a depth of 800' and a 'back vein' was opened up. Ore from this vein is now being milled in a 20-stamp mill, the returns from which are said to be more than paying the expenses of development. The veins occur in amphibolitic schist.

The ore in the various veins is composed of quartz and altered wall-rock, carrying coarse and fine free gold, and 2% of sulphides, pyrite, galena and zinc-blende. In the Tungsten vein, there is a stringer of scheelite varying from 2" to 7" in width associated with the quartz, and free gold was found in the scheelite. The Union Hill veins are credited with a production variously estimated at from $500,000 to $750,000.

The mine is equipped with hoist, new Sullivan air compressor (24" x 14½" x 18"), driven by 200-horsepower motor; Cornish pump driven by a 12' water wheel under 600' head, and a 20-stamp mill driven by a 50-horsepower motor.

The development work done in the last few years has evidently been confined to a 'barren zone' such as has been encountered in every mine in the Grass Valley district. In the Empire mine such a zone was encountered at the 1300' and extended to the 2100' level, from which area practically no ore was recovered. Systematic development will in all probability develop other orebodies.

**Valentine Group.** Owner, F. W. Bradley, Crocker Building, San Francisco.

Location: Washington Mining District, Sec. 5, T. 17 N., and Sec. 32, T. 18 N., R. 11 E., 2 miles southwest to Washington City (N. C. N. G. R. R.). Elevation 4000'–4500'.
Bibliography: U. S. Geol. Survey, W. Lindgren, Prof. Paper 73, pages 139–141. U. S. Geol. Survey Folio 66, Colfax.

The Valentine property consists of 2 locations: the Valentine No. 1 and Valentine No. 2, a total of 40 acres, covering 3000' along the lode. The property is situated on the south slope of the ridge which runs from Washington to Gaston, dividing Poorman and Canyon creeks. There is good timber on the property. Very little work has been done on these claims.

**Vulcan and Grey Eagle Group.** Owner, L. F. Tuttle, 350 Russ Building, San Francisco.

Location: Rough and Ready Mining District, Sec. 6, T. 15 N., R. 8 E., 4 miles southwest of Grass Valley, by fair road. Elevation 2000'.
Bibliography: U. S. Geol. Survey, W. Lindgren, Prof. Paper 73, pages 121–125. U. S. Geol. Survey Folio 18, Smartsville.

This property embraces 2 patented claims: the Vulcan and Grey Eagle, having a total area of 40 acres and a length along the lodge of 1500'. The surface is a rolling plateau. There are other claims located, but they are said to be open, as no assessment work has been done.

The deposit is made up of a quartz vein and stringers, supposed to be a southern extension of the West Point vein system. It varies from a few inches to 1' in width and strikes north. Developed by shallow prospecting only. The ore is free milling and contains pyrite. Country rock is gabbro-diorite and amphibolite.

Adjoining properties are the Kenosha, West Point and Normandie.

**Washington Mine** (German). Owner, Washington Mining Company. Bonded by Von Schroeder Investment Company, 743 Capp street, San Francisco; Baron Von Schroeder, president, San Rafael; H. Von Cleve, secretary.

Location: Washington Mining District, Sec. 9, T. 17 N., R. 11 E., 22½ miles southwest to Nevada City (N. C. N. G. R. R.), by road. Elevation 2800'.
Bibliography: Cal. State Min. Bur. Rept. XII, page 244. U. S. Geol. Survey, W. Lindgren, Prof. Paper 73, pages 139–141. U. S. Geol. Survey Folio 66, Colfax.

The Washington property includes the following claims: Dal, Dee, Don, Becker, Baron fraction, Ocean Star, German mill site, German and German S. Extension. There is a total area of about 480 acres of timber and mineral ground, which covers a length along the lode of 1 mile. This property has been held by the present owners for the past 25 years, but was last worked 18 years ago.

There are several small veins in slate carrying free gold, and less than 1% of sulphides. The average width is 4', the strike north and the dip 85° E. to vertical. From a 400' shaft on the vein, 150' above the river, 2000' to 3000' of stopes have been run on different levels.

There is an old 20-stamp mill on the property, the iron work being in good condition. Also a steam hoist.

Water, under a 30' head, is available from a ditch 1 mile in length. The property was bonded to Von Schroeder Investment Company in 1916, but was idle in September, 1918.

**Washington Chief Claim.** Owner, H. P. Stow, Forbestown, California.

Location: Washington Mining District, Sec. 36, T. 18 N., R. 11 E., 7 miles east of Washington, thence 19 miles southwest by road to Nevada City (N. C. N. G. R. R.). Elevation 4500'.
Bibliography: U. S. Geol. Survey, W. Lindgren, Prof. Paper 73, pages 139–141. U. S. Geol. Survey Folio 66, Colfax.

**West Point and California Mines Company.** (See also **California Mine**, Pittsburg.) Owner, California Mining Company, Grass Valley; K. C. Dunn, agent, 1154 O'Farrell street, San Francisco.

Location: Rough and Ready District, Sec. 31, T. 16 N., R. 8 E., 3½ miles west of Grass Valley. Elevation 2400'.

This mine is situated in the Deadman Flat district, about one-half mile north of the Kenosha, and includes two claims containing about 25 acres of patented land, which have a length along the lode of 2500 feet. The veins strike S. 35° W. and dip 65° NW., occurring in a complex of massive amphibolite, diorite and gabbro, in which the ferro magnesian minerals have been altered near the surface, giving the wall-rocks a reddish brown sandy appearance. There are two ledges 40 feet apart, which may be branches of the same fissure with a horse of country rock between. The veins both average about 1' in width and the filling consists of quartz and altered wall rock. The ore contains heavy free gold with from 5 to 10 per cent of sulphides consisting of pyrite, galena, and chalcopyrite. It was worked at intervals to a depth of 240' by an inclined shaft and produced in 1875 $20,000 from 250 tons of rock, and in 1891 $16,000. The property was again worked in 1910 by a local corporation, who drove a crosscut from the 200' station, 40' into the hanging wall, where a vein was encountered and drifts were driven on the vein 300' north and 200' south of the crosscut. An orebody 170' in length was developed and some of the ore was crushed in the Kenosha mill, but the returns are unknown. The property was equipped with a 35-horsepower compressor and 15-horsepower Cornish pump, transformers, buildings, etc., but this equipment is at present of little account. Electric power was obtained from the Kenosha power line. The gulches in the vicinity of the California, Dulmaine and Kenosha veins were very rich in early days with heavy gold which was in part derived from these lodes, and the showing so far made undoubtedly warrants further systematic development work on the California vein. Work was resumed in November, 1912. J. W. Degelier took a bond on the property in January, 1915. Consolidated in April, 1915, with the West Point mine of the Miller claims into the West Point and California Mines Company, San Francisco. Edward S. De Goyler, president, and T. Cunningham, secretary. The surface plant, including mill, hoist and head frame, was destroyed by fire in June, 1915.

**Wetteran Ranch.**  (See under Copper.)

**Willow Valley.**  (See Montana.)

**Wisconsin.**  (See Empire West Mines.)

**W. Y. O. D.**  (See Pennsylvania.)

**Yellow Metal Group** (Jeffersonian).  Owner, Lakemont Gold Mining Company, Hollidaysburg, Pennsylvania; H. H. Jack, Hollidaysburg, Pennsylvania; F. J. Over, Secretary, Hollidaysburg, Pennsylvania.

> Location: Graniteville Mining District, Sec. 9, T. 18 N., R. 12 E., 7 miles northeast of Graniteville, about 15 miles north of Emigrant Gap (S. P. R. R.) by road.  Elevation 5350'.
> Bibliography: U. S. Geol. Survey, W. Lindgren, Prof. Paper 73, page 141.  U. S. Geol. Survey, Folio 66, Colfax.

The Yellow Metal property comprises 5 claims, the Yellow Metal, Golden Rod, Dollarsign, Beth and Beth Extension, a total of 90 acres, covering a length along the lode of 4500'.

This property was located in 1907 by A. D. Keller et al. H. H. Jack bought the controlling interest. Assessment work has been done each year, and the claims were patented in 1914.

The lode is developed by open cuts only. A 250' crosscut tunnel, if driven 100' further, will probably cut the vein and give 98' of backs.

The deposit is a brown porphyry dike of mineralized stockwork shot full of quartz stringers and kidneys, the whole mass carrying free gold and 3% of pyrite, and galena. The foot-wall is quartz porphyry, distinct from the vein porphyry. No hanging wall has been developed, but it is probably limestone of the Calaveras formation. The dike is 40' in width, strikes north along the contact and dips 30° E. 500' of the mineralized zone has been proven on the surface, and the dike has been traced 1500' northerly to where it passes under Bowman Lake. A mill test run on 3 tons, quartered down to 350 lbs., plated $6.80 in free gold, and showed $4 in sulphides. There are about 5000 tons of ore from open cuts on the dump.

Power is available from the Pacific Gas and Electric Company, 3 miles away, or a 2 mile ditch would give water under 250' head.

There is a boarding house on the property.

**Yerba Buena.** Owners, Sherman W. Marsh, Nevada City; H. Brand. Area 40 acres (patented). No work has been done in late years.

**Yuba Consolidated Gold Mining Company.** Owner, Yuba Consolidated Gold Mining Company, 112 Market street, San Francisco; J. H. Hunt, president; E. L. Gould, secretary.

Location: Washington Mining District, Secs. 11, 14 and 15, T. 17 N., R. 11 E., 5 miles east of Washington, thence 19 miles by road southwest to Nevada City (N. C. N. G. R. R.), 12 miles by road to Emigrant Gap (S. P. R. R.). The Nevada City road is kept open the year round, the Emigrant Gap road is closed throughout the winter. Elevation 3000'-4500'.
Bibliography: U. S. Geol. Survey, W. Lindgren, Prof. Paper 73. U. S. Geol. Survey Folio 66, Colfax.

The Yuba Consolidated Gold Mining Company's property consists of 16 mining claims, one mill site, 3 water rights and 440 acres of timber land, in two groups, the Yuba and Grey Eagle. Eight of the mining claims, the millsite and the timber land are patented. In the Yuba group the Yuba Nos. 1 and 3 and the Schoolhouse claims, each consisting of 20 acres, are not patented. The patented properties of this group are the Yuba No. 2 (16.91A), Mayflower (8.10A), Jokkaree (7.16A), Hathaway (4.50A), Salathiel (50.11A), Yuba mill site (2.50A), and the timber land (440A). Three of the claims of the Grey Eagle group are patented. The claims lie on both sides of the Yuba River.

Former owners produced about $2,000,000 in gold from the Yuba claim, on which were the main workings.

The principal development has been done on the Yuba and Mayflower claims of the Yuba group, and the Grey Eagle, Central and Blue Jay claims, the work consisting of two tunnels, in both of which ore is exposed. On the Yuba claim a tunnel has been driven into the mountain a distance of 1600 feet. At 600' from the mouth a shaft was sunk to a depth of 960' and levels have been run in the vein both north and south at each 100 feet. There are three known veins, the Main, West and East veins, all proven. Nearly all work has been done on the Main vein, south of the point where it is cut by a 4' dike. The Salathiel was worked some years ago, but all of the mine workings are now closed and caved in and one can see only the depression and caving along the line of the tunnel for 400'-500', showing the course of the old stopes. On the Grey Eagle group there are 9 known veins of ore, all of which have been proven to a certain extent. Five have been thoroughly proven, and average from 4' to 14' in width. There are numerous tunnels, drifts and crosscuts, shafts, winzes and upraises in the workings.

The formation is granodiorite with a heavy slate filling along the veins, the lode lying about one-half mile east of the slate belt. The veins course nearly north and south with a dip of from 55° to 70° E. The walls are hard, the ore coming away clean. The Yuba vein varied in width from 3' to 15' and the pay shoot, worked 800' long to a depth of 800′, yielded about $200,000. The Mayflower ledge would mill from $5 to $10 per ton. The filling in the veins is quartz, and from ore milled, it is estimated that all of the ore in the property will return from $6 to $8 per ton.

There is a 10-stamp mill at the Grey Eagle and a 20-stamp mill at the Yuba, 15 of which are in order and will last for some time; also a 4-drill compressor and a water power hoist, the latter being in good condition. A Dow pump of 150 gallons capacity will suffice, if supplemented by a station pump at the 300′ level.

Water is conveyed by two flumes; one about one-half mile up stream, the other one mile, the rights being among the oldest on the river. The river can be relied on for water power for 9 months of the year, but during the summer, the flow in dry years might not exceed 500″. The head at the compressor is 170′.

There is a saw mill on the timber property with a daily capacity of from 6 M. to 10 M. feet.

**Zebright Mine.** Owner, Bear Valley Mining Company, 1200 Battery street, San Francisco. Under bond to F. H. Turner, Emigrant Gap.

Location: Lowell Hill Mining District, about 27 miles from Nevada City and about 5 miles from Emigrant Gap (S. P. R. R.).
Bibliography: U. S. Geol. Survey, W. Lindgren, Prof. Paper 73, pages 146–147. U. S. Geol. Survey Folio 66, Colfax.

The Zebright mine consists of two locations, the Zebright and Zebright Extension, on the north side of Bear River.

A tunnel has been driven from the mill, a distance of 990 feet to the north, giving about 400′ of backs on the dip of the pay shoot. Another tunnel 96′ lower than the present working tunnel has been driven to a point about 100′ from the first pay shoot. The rock from the lower tunnel would have to be hoisted to the mill, but there is a hoist in place with water power and cable. The main shoot has been stoped, following its dip, for a distance of about 400 feet. The length on the vein is about 160′.

The country rock is slate, metamorphosed on the hanging wall and on the foot-wall, slightly changed at contact with the ledge. The vein is nearly perpendicular and averages about 20′ in width, being in the nature of a dike extending across the country for many miles. The ore ranges from $2.10 to $3 per ton and it was being mined and milled for $1.70 per ton. The sulphurets are worth from $12 to $14 per ton. Gross production daily is about $75.

The reduction plant consists of a 10″ x 12″ jaw crusher, 10 1150-lb. stamps and a 4′ ball mill. The compressor handled 3 Water-Leyner drills and 2 stopers. A pipe line extends from the South Yuba ditch down to the quartz mill and compressor, giving a water pressure of

470'. An 8' overshot water wheel furnishes power for the mill which handled from 25 to 40 tons of ore per day. The battery discharges into a classifier. The fines pass directly over the amalgamation tables while the coarse material is automatically fed into the ball mill which discharges on to an amalgamation table.

The ore is hand-sorted and about one-third of the material worked is waste.

### GOLD—PLACER MINES.

**Alta-California.** (See under Drift.)

**Bigg and Sims.** (See under Drift.)

**Birdseye Creek.** (See You Bet Mining Company, Drift.)

**Blue Lead.** (See under Drift.)

**Cold Springs.** (See under Drift.)

**Delaware.** (See under Drift.)

**Derbec.** (See under Drift.)

**Eagle Echo.** (See under Drift.)

**Fountain Head.** (See under Drift.)

**Grover.** (See under Drift.)

**Harmony.** (See under Drift.)

**Hirschmann.** (See Grover, Drift.)

**I. X. L.** (See under Hydraulic.)

**Jenny Lind.** (Seen under Drift.)

**Kate Hayes.** (See River Mines Company, Hydraulic.)

**Le Duc.** (See under Lode.)

**Lupine.** (See under Drift.)

**Major.** (See under Drift.)

**Manzanita.** (See under Drift.)

**Murphy.** (See Grover, Drift.)

**Nichols.** (See under Lode.)

**North Bloomfield.** (See under Hydraulic.)

**Odin.** (See Manzanita, Drift.)

**Omega.** (See under Hydraulic.)

**Oriental.** (See Deadwood, Lode.)

**River Mines Company.** (See under Hydraulic.)

**Sazarack.** (See under Drift.)

**Twin Sisters.** Owner, Chas. Hagerty, Bloomfield.

**West Harmony.** (See Delaware, Drift.)

## MAGNESITE.

An outcrop of magnesite three feet wide is said to occur on the Whilden and Bishop property, 6 miles southwest of Graniteville.

## MANGANESE.

E. H. and W. E. Bartholf and J. A. Veach, Box 97, Colfax, were mining manganese in June, 1918, 7 miles north of Colfax and 5 miles from Bear River bridge on the west side of the river. One car of ore was shipped from here three years ago and two more carloads were said to have been mined and ready to ship late in June.

The **Wren Ranch** property consists of 160 acres in Sec. 20, T. 14 N., R. 8 E., M. D. M., in Lime Kiln District, at an elevation of 1000 feet. It is owned by R. M. Wren of Wolf, Nevada County, and leased to C. E. Loufbourrow of Oakland.

A body of manganese 14″ thick and 10′ long has been exposed to a depth of 20 feet by an inclined shaft. Ten tons of ore mined up to November, 1917, is said to assay 50% to 60% metallic manganese and 6% silica.

## PYRITE.

**Iron Mountain Deposit** (Taylor Iron Mine). Owners, Mammoth Copper Company, Kennett, Shasta County, California; U. S. Smelting and Refining Company, Hobart Building, San Francisco.

Location: Indian Springs Mining District, Sec. 4, T. 15 N., R. 7 E., 10 miles southwest of Grass Valley. Elevation 1500′–1900′.

This deposit has been known for many years as an iron mine on account of its extensive gossan or iron capping. No attempt, however, was made to develop the property until the present owners started work in 1906. This company now owns 300 acres of mineral rights under the Robinson & Sonford agricultural holdings. These rights cover about 4000 feet along the strike of the deposit.

Between 1906 and 1908 a vertical shaft was sunk to a depth of 265 feet and levels were run at depths of 125, 150 and 250 feet. On the 250′ level, below the iron cap or oxidized zone, a crosscut was driven from the shaft 125′ to the rhyolite foot-wall, and along this wall drifts were extended 300′ and 400′ west of the crosscut. From these drifts intermediate crosscuts were driven from the foot-wall to the hanging wall at intervals of 50 feet. This work proved the sulphide zone to be from 500 to 600 feet in length and to vary in width from 125 feet at the center to a few feet at each end. According to Mr. Henry Schroeder, who was superintendent of the property, the hanging wall is diabase and there is from 1′ to 7′ of talc on the foot-wall which carries gold. The so-called sulphide orebody consists of pyrite which for a distance of 40 to 50 feet from the foot-wall is said to carry 1½% of copper. The main sulphide body carries 50% sulphur, but very little copper. Mr. Schroeder is of the opinion that similar sulphide bodies will be developed on this property. The mine was closed in 1908, and all the equipment with the exception of an 80-horsepower boiler was removed.

**Spence Mineral Company.** (See Under Copper.)

## TUNGSTEN.

**Union Hill Mine.** (See also under Gold.) The Tungsten vein was found in a crosscut driven 500 feet south from the 600-foot level of the Union Hill Shaft. Considerable drifting and stoping has been done on this vein, a large part of the ore mined and milled in the past three years having been worked for its scheelite content. On the 600-foot level the Tungsten vein strikes N. 55° W. and dips 70° SE. The vein is 6 feet wide between slate walls and the ore occurs in seams from one-half inch to 16 inches wide, with the scheelite tending to follow the foot-wall. The vein carries sulphides and some free gold is associated with the scheelite. On the 800-foot level this vein unites with another. It strikes N. 45° W. and dips 65° S.E. The vein is 16″ wide and the scheelite still tends to follow the foot-wall.

The scheelite ore was crushed to 8-mesh in 5 stamps of the mill. The pulp passed over a large size Deister Simplex Sand Concentrator on which coarse and fine scheelite was caught. The coarse tailing passed to a hopper and thence to 5 stamps with 20-mesh screen, being concentrated on another Deister table. The tailing from this table is concentrated on a third of similar type. This last concentrate carried about 12% scheelite and most of the pyrite. It was sweet roasted to oxidize the pyrite and free the gold and was brought back for retreatment to save the gold and scheelite separately. The slimes from the ten stamps were passed over plates, then into Deister precipitating cones and finally over a Deister Simplex Slime Concentrator.

The Union Hill mine was one of those affected by the miners' strike in June, 1919, and is said to have been closed indefinitely early in July.

# INDEX.

www.ingramcontent.com/pod-product-compliance
Lightning Source LLC
Chambersburg PA
CBHW060337200326
41519CB00011BA/1960